NATURE AND NATURE'S LAWS

Themes in science and literature

by

Jack Meadows

with

Sally Hodges

©2015

NATURE AND NATURE'S LAWS

Themes in science and literature

By Jack Meadows with Sally Hodges

Copyright ©Jack Meadows 2015

Front cover by Caz Woolley

CONTENTS

Then comes Mr. Malthus forward with the geometrical and arithmetical ratios in his hands, and holds them out to his affrighted contemporaries as the only means of salvation.
[William Hazlitt *The spirit of the age*]

"Let x denote beauty, - y, manners well-bred,
z, fortune - (this last is essential),
Let L stand for love" - our philospher said,
"Then L is a function of x, y, and z,
Of the kind which is known as potential."
[W.J.M. Rankine *The mathematician in love*]

Such are a very few of the considerations which show that for every type of animal there is an optimum size. Yet although Galileo demonstrated the contrary more than three hundred years ago, people still believe that if a flea were as large as a man it could jump a thousand feet into the air. As a matter of fact the height to which an animal can jump is more nearly independent of its size than proportional to it.
[J.B.S. Haldane *On being the right size*]

Earth goes by chemic forces; Heaven's
A Mécanique Céleste!
And heart and mind of human kind
A watch-work as the rest!
[A.H. Clough *When Israel came out of Egypt*]

An inextensible heavy chain
Lies on a smooth horizontal plane,
An impulsive force is applied at A,
Required the initial motion of K.
[James Clerk Maxwell *A problem in dynamics*]

Big whirls have little whirls
That feed on their velocity;
And little whirls have lesser whirls,
And so on to viscosity.
[L. F. Richardson *Turbulence*]

In aerodynamics, because you've only got the thing on paper at first, you use dimensionless coefficients: ratios of this to that - centimeters, grams, seconds neatly all cancelling out above and below. This allows you to use models, arrange an airflow to measure what you're interested in, and scale the wind-tunnel results all the way up to reality, without running into too many unknowns, because these coefficients are good for all dimensions. Traditionally they are named after people - Reynolds, Prandtl, Péclet, Nusselt, Mach.
[Thomas Pynchon *Gravity's rainbow*]

And find for one small world of fact
Invariant matrices, compact
Within the dark and igneous rock
Of Comptes Rendus *or* Proc. Roy. Soc.
[Michael Roberts *Note on θ, φ and ψ*]

Series of numbers occur in most areas where mathematics is applied, not least in economics. Robert Malthus, a leading political economist in the early years of the nineteenth century, used his knowledge of number series as the basis for calculations on population growth. He argued that increases in the amount of food produced can only grow at an arithmetic rate, whereas human beings can reproduce at the much faster geometric rate. This led him to his much-debated conclusion that the population must always eventually exceed its supply of food. Hazlitt, a follower of Rousseau, strongly criticised this gloomy conclusion [first extract]. The nineteenth-century Scottish engineer, William Rankine, joked about the possibility of applying mathematics to human activities – a matter for discussion at the time [second extract]. Rankine was a pioneer of the theory of steam engines, and emphasized the role of 'potential energy', which appears in the stanza quoted here.

Galileo described what is sometimes called the square-cube law in his book *Two new sciences* of 1638. This simply says that, as a body grows in size, its volume grows faster than its area. As applied to animals [third extract], this implies they get relatively weaker as they grow in size, since the cross-section of their muscles grows more slowly than their volume-related weight. Haldane's Marxist beliefs encouraged him to spend much time on the popularisation of science, often using simple quantitative arguments.

Laplace was one of the leading applied mathematicians at the end of the eighteenth century. His discussion of how bodies in the solar system moved developed and extended the studies that Newton had begun a century before. Laplace's results were published in a series of volumes entitled *Mécanique Céleste*. The point of Clough's comments [fourth extract] was this. Newton had seen the solar system as so complex that it would sometimes require external correction. Laplace claimed that the solar system was stable: there was no need to invoke God's assistance.

Nineteenth-century mathematics in Britain was dominated by the Cambridge tripos examinations. An essential component of these was the study of bodies in motion. Maxwell both sat these exams as a student and later set questions as professor at Cambridge. His poem [fifth extract] reflects how such questions were phrased.

The study of how solid bodies move is hard enough, but the study of fluid motion is even harder. Richardson was the first to try and predict the weather by numerical calculation in the 1920s. A major problem in trying to deal with fluid flow - as in the atmosphere - was the presence of turbulent motion, since it proved difficult to devise a mathematical theory of turbulence. He summarised the nature of turbulence in rhyme [sixth extract] - a parody of Swift's verse concerning greater fleas having lesser fleas.

As Pynchon says [seventh extract], fluid mechanics makes great use of dimensionless numbers, because they are independent of scale. These numbers are typically called after people who carried out work relevant to them. The Reynolds number, for example, indicates whether flow is smooth or turbulent, and is called after the British physicist, Osborne Reynolds. The Mach number relates the flow velocity to the velocity of sound in the fluid, and is called after Ernst Mach, an Austrian physicist. Prandtl and Nusselt were Germans; Péclet was French. The list is not comprehensive: there is also, for example, a Richardson number.

A matrix is an array of numbers or symbols, arranged in rows and columns, which can be manipulated according to a set of rules [final extract]. An invariant matrix is one that does not change - for example with time. Matrices are used in a variety of physics-related applications, including quantum mechanics, fluid flow and geophysics. Greek letters are, of course, common symbols in mathematics. The *Comptes Rendus* of the French Académie des Sciences and the *Proc. Roy. Soc.* [*Proceedings of the Royal Society of London*] are two of the oldest outlets for publishing scientific research. They are, perhaps, 'dark' because of their complicated contents and 'igneous' because their originally malleable contents are set in stone after publication. Michael Roberts trained as a mathematician at Cambridge in the 1920s. He subsequently became a schoolteacher and poet.

The atomic bombs had dwarfed the international issues to complete insignificance. When our minds wandered from the preoccupations of our immediate needs, we speculated upon the possibility of stopping the use of these frightful explosives before the world was utterly destroyed. For to us it seemed quite plain that these bombs and the still greater power of destruction of which they were the precursors might quite easily shatter every relationship and institution of mankind.
[H.G. Wells *The world set free*]

Throughout the ages--and long before the invention and development of nuclear weapons--there had been those who prophesied that the world would end because of man's wickedness. Such prophesies were always believed, no matter how many times they had been proved wrong in the past. There was a wish for, as well as a fear of, punishment. Once nuclear weapons were invented, the prophecies gained plausibility, although now they were couched in lay terms rather than religious ones.
[Brian Aldiss *Helliconia Winter*]

Give an embrace that will bring back
Tears of good to everyone's heart
Then spring at them all over the world
Shouting, "We are the boys and girls,
The Children of Hiroshima!"
[Toge Sankichi *At the First Aid Station*]

Otto Hahn wants to kill himself, because it was he who discovered fission, and he can see the blood on his hands. Gerlach, our old Nazi co-ordinator, also wants to die, because his hands are so shamefully clean. You've done it though. You've built the bomb.
[Michael Frayn *Copenhagen*]

They made a myth of you, professor,
*   you of the gentle voice,*
*   the books, the specs .....*
Now it's 'Mr. Atilla, how do you do?'
Do you pack wallops of wholesale death?
[Carl Sandburg *August, 1945*]

They asked me what I thought of the atomic bomb. I said I had not been able to take any interest in it. I like to read detective and mystery stories. I never get enough of them but whenever one of them is or was about death rays and atomic bombs I never could read them.
[Gertrude Stein *Reflection on the atomic bomb*]

In the Balintang Channel he found much dust and radioactivity far above the lethal level. On the seventh day of the war he was in Manila Bay ... The atmospheric radioactivity was rather less here, though still above the danger level. [Nevil Shute *On the beach*]

I yell thru Washington, South Carolina, Colorado,
*   Texas, Iowa, New Mexico,*
Where nuclear reactors create a new Thing under the
*   Sun, where Rockwell war-plants fabricate this death*
*   stuff trigger in nitrogen baths*
[Allen Ginsberg *Plutonian ode*]

Atomic bomb   DISCUSSION

The first few decades of the twentieth century saw a remarkably rapid growth in our understanding of atoms and of their internal structure. By the outbreak of the First World War, it was known that all atoms consist of a small central nucleus surrounded by a cloud of electrons. It was also known that the nuclei of the heaviest atoms, such as Uranium, gradually break down over time releasing appreciable energy in the process. (This was called 'radioactivity' - an unfortunate choice of name, since the phenomenon has nothing to do with radio.) It was popularly speculated that these discoveries might lead to a new type of weapon. Wells' novel [first extract], published in 1914, gave voice to these speculations. (His supposed bombs were not particularly powerful, but they continued to explode over a long period of time.)

Wells' usage helped popularise the term 'atomic bomb', which is still widespread today. This is again a pity, since, the energy of real bombs actually comes specifically from the nucleus. Consequently, we should be talking about 'nuclear weapons' [second extract]. Once it became apparent, during the Second World War, that such weapons could be made to work, there followed the question of their effects. For example, could the explosion of a hydrogen fusion bomb under water ignite the hydrogen in the water itself? At a less abstruse level, what might the explosion of a series of atomic bombs do in terms of polluting the Earth's atmosphere? The only direct data on the effects of atomic bombs have come from studies of the two Japanese cities, Hiroshima and Nagasaki, which were bombed in 1945. The 24-year old Toge Sankichi was living in Hiroshima when the bomb fell there. Although he was already writing poetry, he is best known for the poems he wrote in the wake of that event [third extract]. He was an early contributor to what subsequently became a new literary genre in Japan - atomic bomb literature. Toge Sankichi, himself, died of leukaemia at the age of thirty-six.

The crucial event in the decision to build an atomic bomb was a discovery made not long before the Second World War began. This was that the nuclei of heavy elements can be split into fragments, if they are bombarded by neutrons. ['Neutrons' are electrically neutral particles found in atomic nuclei.] Such splitting releases both energy, and a number of new neutrons. These latter can then fragment further nuclei, leading to a chain reaction and the explosive production of large amounts of energy. Michael Frayn's play concentrates on a meeting between the German physicist, Werner Heisenberg, and the Danish physicist, Niels Bohr, in which they discuss the creation of the atomic bomb. Hahn and Gerlach [fourth extract], both leading German scientists, were interned briefly by the British at the end of the Second World War as part of an investigation of German attempts to develop a bomb, and their conversations were recorded. Otto Hahn played a leading part in the discovery of nuclear fission, but was horrified to hear of the creation of the bomb. Walter Gerlach was more inclined to be sorry that Germany had failed to make a bomb before the War ended. The Second World War has sometimes been referred to as the 'physicists' war'. The reason is that it saw the development of two important inventions based on physics - radar and the atomic bomb. As a consequence, the popular image of the scientists involved changed. Sandburg's poem [fifth extract] reflects this. (The first atomic bomb was dropped on Hiroshima on 6 August 1945, and the second on Nagasaki on 9 August.)

However, early reactions to the bomb were very varied. Gertrude Stein [sixth extract] seems to have been mainly bored by it. She saw it as yet another wearisome apocalyptic threat: perhaps she was reacting against the sort of attitude noted in the second extract. At the other end of the scale, was the Armageddon scenario portrayed in Nevil Shute's novel [seventh extract]. Originally published in 1957, it is still in print and has been adapted for films, television and radio. It envisaged a nuclear war in the Northern hemisphere, with the lethal fallout gradually moving southwards to cover Australia - the geographical focus of the plot - and killing off all mankind. Fairly soon after the War, protests began against the further development of the atomic bomb and the proliferation of nuclear weapons.

The American Beat poet, Allen Ginsberg, was one of the protestors. In particular, he tried to blockade the transport of fissile material to the Rockwell Corporation plant in Colorado, where bomb components were being manufactured. This was in 1978, the year in which his *Plutonian ode* was written [eighth extract]. Activities at Rockwell included the use of radioactive plutonium, but Ginsberg's use of the adjective 'Plutonian' suggests that he was also thinking of Pluto, the god of the underworld in classical mythology.

..... saw the flaring atom-streams
And torrents of her myriad universe,
Ruining along the illimitable inane,
Fly on to clash together again, and make
Another and another frame of things
For ever.
[Alfred Tennyson *Lucretius*]

Atom — the atoms in the etherial substance, how by fermentation they form a hard body, supposed by Epicurus to be indivisible, how their simple motion in a fluid confined in a small space causes hardness, the hard atoms in the air have each a very swift simple motion, differ in consistence, figure, motion, and magnitude
[Thomas Hobbes *Index to his Collected Works*]

The atoms of Democritus
And Newton's particles of light
Are sands upon the Red Sea shore
Where Israel's tents do shine so bright.
[William Blake *Mock on, mock on! Voltaire, Rousseau*]

At quite uncertain times and places,
The atoms left their heavenly path,
And by fortuitous embraces,
Engendered all that being hath.
[James Clerk Maxwell *Molecular evolution*]

Around the nucleus (Bohr says)
Fierce spots of energy,
Upon existence edge,
Are spun
In discrete orbit each.
[H. Witheford *Bohr on the atom*]

The atom is a crystal
of a sort; the lattices
its interlockings form
lend a planarity most pleasing
to the abysses and cliffs, much magnified,
of (for example) salt and tourmaline.
[John Updike *Ode to crystallization*]

                    Now eyes, raised
To the nth power by the power of minds
over matter, can trace the medallions the little
particles make in the miniscule nets
of lace that lattice every least tittle
of space in the tin in the point of a pin.
[Dorothy Donnelly *The point of a pin*]

The idea that the world consists solely of atoms and void appeared several centuries BC. Its best known formulation is in the poem *De rerum natura* [Concerning the nature of things] written by the Roman poet Lucretius in the first century BC. His poem was designed to explain the ideas of the Greek philosopher, Epicurus, from some 250 years before. Tennyson's poem [first extract] succinctly summarises what Lucretius had to say about atoms. Atoms came in various shapes and sizes and were in continual motion. As they clashed together, they formed a variety of objects which, in due course, broke up, only to aggregate into new objects, and so on for ever. (The Greek word from which our word 'atom' derives means 'uncuttable' or 'indivisible'.)

Aristotle believed, on the contrary, that matter was continuous, and it was his view that was accepted during the European Renaissance. From the beginning of the seventeenth century, however, the idea of atoms came to be increasingly discussed: by prominent: figures in the history of science: Robert Boyle, for example, supported it. The problem was that atomism traditionally was linked to atheism - something that the natural philosophers of the day wished to avoid. It did not help that Thomas Hobbes (who was widely believed to be an atheist) supported atomism. As the second extract illustrates, he was acquainted with all the basic ideas described by Lucretius. Fortunately, his views were something closer to the picture put forward by the French philosopher, Descartes. This helped, for few held to the Cartesian world view in Britain (and it was refuted by Newton in his *Principia* later in the century).

Early in the eighteenth century, Newton published his book, *Opticks*, in which he supported the belief that light consists of particles (or corpuscles as they were often called) and this remained the standard picture of light for the next hundred years. William Blake continued to associate atomism, and so Newton's light corpuscles, with atheism [third extract]. He goes back beyond Epicurus and Lucretius to start his condemnation with Epicurus' predecessor, Democritus.

It was a basic assumption of atomism that atoms could associate together to form groupings with new properties. The nineteenth century saw major progress in explaining how this worked. On the one hand, it was realised that the world contained a limited number of atom types - called elements - from which everything else was built. On the other, it was realised that the build up consisted of atoms coming together in clearly defined ways to form molecules. These molecules could, in turn, join together to form larger molecules. This is why Maxwell entitled his poem *Molecular evolution* [fourth extract]. The reason why atoms join together in only a restricted number of ways was explained in the twentieth century, when the internal structure of atoms was explored. The earliest theoretical model for the interior operations of an atom was put forward by the Danish physicist, Niels Bohr, just before the First World War [fifth extract].

Twentieth-century study of atoms also explained why differently shaped crystals occur: it depends on the way that molecules of different kinds attract each other to provide a recurring structure. For example, salt, which John Updike mentions, is sodium chloride. The sodium has a positive electric charge and the chlorine a negative charge. Salt crystals form on the basis of the electrical attraction that this generates in a recurring sequence of the two ions. Tourmaline is a much more complicated crystal. It is formed from various silicate compounds and can have more than one structure. It is familiar as a semi-precious stone which can occur in a variety of colours.

Optical microscopes cannot be used to see individual atoms because the wavelength of light is much larger than sizes of the atoms themselves. Electron microscopes - ones that use a stream of electrons rather than light to scrutinise an object - can provide much finer detail. At the beginning of the 1980s, a new type of electron microscope - called a scanning tunnelling microscope - was devised. (Its inventors were awarded a Nobel Prize for their work.) This proved capable of seeing individual atoms. Dorothy Donnelly's poem [final extract] appeared not long after machines of this type came into operation. A standard way of viewing individual atoms was to put a specimen sample – tin, in this case - on the end of pin, and to focus the microscope on this point.

..... so work the honeybees,
Creatures that by a rule in nature teach
The act of order to a peopled kingdom.
They have a king and officers of sorts,
Where some like magistrates correct at home,
Others like merchants venture trade abroad,
Others like soldiers armèd in their stings
[William Shakespeare *Henry V*]

Burly, dozing humble-bee,
Where thou art is clime for me.....
Let me chase thy waving lines;
Keep me nearer, me thy hearer,
Singing over shrubs and vines.
[Ralph Waldo Emerson *The humble-bee*]

As I write this, two or three weeks later, I am sitting near the brook under a tulip tree..... it swarms with
myriads of these wild bees, whose loud and steady humming makes an undertone to the whole, and to my
mood and the hour. All of which I will bring to a close by extracting the following verses from Henry A.
Beers's little volume:
As I lay yonder in tall grass
A drunken bumble-bee went past
Delirious with honey toddy.
The golden sash about his body
Scarce kept it in his swollen belly
Distent with honeysuckle jelly.
[Walt Whitman *Bumble-bees*]

The first thing I saw in the morning
Was a huge golden bee ploughing
His burly right shoulder into the belly
Of a sleek yellow pear.
[James Wright *The first days*]

Bees, pure selfless workers,
thin, flashing proletarians, perfect fearsome militia
that in war attack with suicidal stings
buzz, buzz over the earth's realms
family of gold, windy multitudes
[Pablo Neruda *Ode to the bee*]

The rector, the midwife, the sexton, the agent for bees.....
And they are all gloved and covered, why did nobody tell me?
They are smiling and taking out veils tacked to ancient hats.
..... here is the secretary of bees with her white shop smock,
Buttoning the cuffs at my wrists and the slit from my neck to my knees.
[Sylvia Plath *The bee meeting*]

Bees   **DISCUSSION**

Because honey bees live in communities of many thousand, it is tempting to draw parallels between them and human communities. The comparison often dwells on the question of organisation - as Shakespeare does in the first extract. The particular interest of this extract is that the queen bee is referred to as the 'king'. In fact, the queen bee was generally believed to be male at the time. It was only in the latter part of the seventeenth century that the Dutch scientist, Jan Swammerdam, used the newly invented microscope to show that queen bees had ovaries and were therefore female. It is a pity that Shakespeare did not know this, since he would surely have worked in an allusion to Queen Elizabeth as head of the country at this point.

There are many thousand different species of bee distributed all round the world, but humans typically take most notice of two types - the honey bee and the bumblebee. Emerson [second extract] is observing the latter. He calls it the 'humble-bee'. This name dates back to at least the fifteenth century (and was used by Shakespeare). However, the alternative name 'bumblebee' appeared not long after - in the sixteenth century - and the two names continued in parallel use for some centuries. Up to the First World War, 'humble bee' was more popular, but, since the Second World War, it has almost entirely disappeared. 'Humble' refers to the humming noise the bee makes. 'Bumble' refers partly to this, and partly to the weaving path of a bee when it is flying - as Emerson notes. 'Drunken' [third extract] also reflects this, but links on to 'toddy'. This was originally a drink made from palm sap, but the name was subsequently applied to any hot, sweetened alcoholic drink. Flowers produce a sugary nectar, which some bees are able to collect via long tongues. The nectar is mixed with enzymes and stored within the bee, so producing a 'swollen belly'. (Henry Beers was a writer who pioneered the discussion of American literature as a separate entity from British literature.) By the autumn, flowers become scarcer; but over-ripe, or fallen fruit provide a new source of food. Wright's poem [fourth extract] concerns this time of the year in Italy. It describes how the incursion of a bee causes an over-ripe pear to fall, requiring the writer to come to the bee's rescue.

Pablo Neruda was both a poet - he won the Nobel Prize for Literature - and a leading Communist politician in Chile. His poem on the bee [fifth extract] takes us back to the comparison of human and bee communities with which we began. 'Pure, selfless workers' reflect an ideal Communist vision, as does the suicidal defence of one's community. Honey bees are unique in having a strongly barbed stinger, which is difficult to withdraw from the wound made by the sting. Leaving the sting behind in the victim cripples the bee and quickly leads to its death. Bumblebees have smooth stingers, as does the queen in honey bees, so they can sting repeatedly. From a bee's viewpoint, this might seem to be a more rational option than committing suicide.

In 1961, Sylvia Plath and Ted Hughes moved to a village in Devon, and the following summer they took up beekeeping. Plath, in particular, was fascinated by the activity, and conceived a sequence of five poems devoted to bees. She saw this sequence, which tied in with her thoughts about community and self, as a breakthrough in her work as a poet. Her father, Otto, had written a standard text on bumblebees: as a boy in Germany, he had been nicknamed 'Beinen-Konig' [the bee king]. He subsequently became a professor of biology at Boston University, where he remained obsessed with bees. Ted Hughes believed his wife was herself obsessed with her father. His own poem, *The bee god*, begins with the words: 'When you wanted bees I never dreamed/ It meant your Daddy had come up out of the well'.

Sylvia Plath wrote to her mother about attending a meeting of beekeepers in her Devonshire village, and this forms the basis for the extract from her bee poems quoted here. Beekeepers usually wear protective clothing. Bees are attracted by human breath, so, since stings are particularly painful on the face and neck, hats and veils are commonplace. Many beekeepers also wear gloves, though some experienced beekeepers find it easier to work with bare hands. A bee sting left in clothing can incite further attacks, so beekeepers often wear overalls that cover the whole body - Plath's 'white shop smock' was presumably less extensive. The covering is typically light-coloured because bees associate dark colours with predators. *The bee meeting* describes the beginning of the bee season. The last poem in Plath's sequence - *Wintering* - describes the end, when, 'To make up for the honey I've taken/ Tate and Lyle keeps them going'. (Tate & Lyle - two nineteenth-century firms that amalgamated - has long been the main provider of refined sugar in Britain.)

Four months after completing these bee poems, Sylvia Plath committed suicide.

Though I know something about British birds I should have been lost and confused among American birds, of which unhappily I know little or nothing. Colonel Roosevelt not only knew more about American birds than I did about British birds, but he knew about British birds also. What he had lacked was an opportunity of hearing their songs, and you cannot get a knowledge of the songs of birds in any other way than by listening to them.
[Sir Edward Grey *Recreation*]

While the chaffinch sings on the orchard bough
In England – now! .....
And after April, when May follows,
And the whitethroat builds, and all the swallows
[Robert Browning *Home thoughts from abroad*]

In his first spring there he was ' astonished and delighted ' by the richness of the bird-life ; he never knew so many nightingales. He saw herons go over, and a teal. Magpies were common, and he records ten together on September 9, 1881, within twelve miles of Charing Cross…… ' Birds,' he notes, ' care nothing for appropriate surroundings.'
[Edward Thomas *Richard Jefferies*]

A person lately found a young cuckow in a small nest built in a beechen shrub at the upper end of the bostal. By watching in a morning, he soon saw the young bird fed by a pair of hedge-sparrows. The cuckow is but half-fledge; yet the nest will hardly contain him: for his wings hang out, & his tail & body are much compressed, & straightened.
[Gilbert White *The natural history of Selborne*, 1783]

Each hawk or falcon stood in the silver on one leg, the other tucked up inside the apron of its panel, and each was a motionless statue of a knight in armour. They stood gravely in their plumed helmets, spurred and armed…… In those days they used to hood everything they could, even the goshawk and the merlin, which are no longer hooded according to modern practice.
[T.H. White *The sword in the stone*]

When my late friend Dr. Chambres, of Derby, was on the island of Caprea in the bay of Naples, he was informed that great flights of quails annually settle on that island about the beginning of May, in their passage from Africa to Europe. And that they always come when the south-east wind blows, are fatigued when they rest on this island, and are taken in such amazing quantities and sold to the Continent, that the inhabitants pay the bishop his stipend out of the profits arising from the sale of them.
[Erasmus Darwin *Zoonomia*]

The sky
was dramatic with great struggling V's
of geese streaming south, mare's-tails above them .....
a cloud appeared, a cloud of dots
like iron filings which a magnet
underneath the paper undulates.
..... held an identity firm: a flock
of starlings, as much one thing as a rock.
[John Updike *The great scarf of birds*]

Birds   DISCUSSION

Most branches of biology are now mainly studied by professionals. Birds form an exception: amateur input here is still valuable. Keen ornithologists come from a variety of backgrounds. Sir Edward Grey [first extract] combined being a keen observer of bird life with the post of Foreign Secretary in the early years of the twentieth century. (He is best remembered for his comment as the First World War started: 'The lamps are going out all over Europe. We shall not see them lit again in our life-time.') He records here a visit by Theodore Roosevelt, whose period as US President overlapped with Grey's time in office. Roosevelt is often remembered as a hunter of big game, but he was also interested in natural history and conservation.

Browning's thoughts on English birds draw attention to seasonal changes. The chaffinch sings in April because that is the month that the bird starts breeding. The whitethroat starts nesting a little later in May. It is a summer visitor to England, going south of the Sahara in the winter. Swallows come to England in May from even further afield, since they spend the winter in South Africa. The effect of global warming is to make migrants appear somewhat earlier than in the past. So maybe Browning's timings will need adjustment sometime in the future.

Britain was a hospitable place for birds throughout the nineteenth century, even in and around tows [third extract]. (Developments, such as pesticides, which have had a major effect on bird life, have only appeared in the twentieth century.) Writings about nature, often with a particular mention of birdlife, were immensely popular in the decades before the First World War. Jefferies, who died in 1887, was a well-known English author in this category. Edward Thomas, his biographer, is best remembered now as a poet, but was known then as a writer about the countryside. His most quoted poem *Adlestrop* ends by talking of bird song. Thomas' biography was dedicated to W.H. Hudson, another contemporary British nature writer, who gave a special emphasis to birds.

The latter part of the eighteenth century saw the first detailed studies of cuckoos as parasites. Two of the leading investigators were Edward Jenner, best known for introducing vaccination, and Gilbert White [fourth extract]. It is significant that White mentions the sparrow as the host. It seems to be the most easily deceived of all the birds, readily accepting the differently coloured cuckoo's egg as its own. (A *bostal* is a small road leading up a hill.)
Birds of prey have been used for hunting for many centuries. White sets his story of King Arthur in the mediaeval period when falconry was particularly popular in Britain [fifth extract]. (He was himself a falconer and wrote about his attempts to train a hawk.) The birds were graded according to the sort of person who could fly them. Thus the owner of a merlin would be of higher rank than the owner of a goshawk. The idea of putting a hood on such a bird was to make sure that the first thing it saw after the hood was removed was the designated quarry.

The common quail (there are various species), unlike most game birds, is migratory. Quails have always been used for food. They were a popular part of the menu in Ancient Egypt. (When the Israelites fled Egypt, one of their sources of sustenance was migrating quails.) As Erasmus Darwin notes, quails have long been captured as they pass through the Mediterranean [sixth extract]. Hardly surprisingly, such continued hunting has affected their overall number, though they are not as yet in any danger of extinction.

Although other migrating birds may use a V-formation, it is a particularly memorable sight with geese [seventh extract]. The rationale for using such a formation is aerodynamic. All the birds apart from the leader fly in the upwash from the wingtips of the bird ahead, and this helps to support their weight. It has been estimated that the assistance provided in this way can extend the birds' migratory range by almost three-quarters. The order in which the birds fly is changed at intervals to spread the work load. (*Mare's-tails* is a popular name for cirrus clouds.) A flock of starlings - delightfully known as a murmuration of starlings – provides a remarkable display of coordinated flying [final extract]. Thousands of starlings wheel around at sunset forming black cloud patterns against the sky. The interesting question is how the coordination occurs. Studies in recent years show that each starling keeps a certain distance from its nearest neighbours, so that a change of direction or speed propagates rapidly through the flock. This type of collective behaviour is well known in the scientific world: it occurs in inorganic assemblages, as well as organic. Unfortunately, the starling population in the UK has plummeted in recent years, so this sight has become much rarer.

A speck that would have been beneath my sight
On any but a paper sheet so white .....
This was no dust speck by my breathing blown,
But unmistakably a living mite
With inclinations it could call its own.
[Robert Frost *A considerable speck*]

[I] turned to the rubric of the first known owner dated 1221, the rubric a squiggle of very thick ink. I put a glass on it and there imbedded deep in the ink was the finest crab louse, pfithira pulus, I ever saw. He was perfectly preserved even to his little claws. I knew I would find him sooner or later because people of that period were deeply troubled with lice and other little beasties - hence the plagues.
[John Steinbeck *Letter, September 1962*]

Ye ugly, creepin' blastit wonner,
Detested, shunn'd by saunt an' sinner,
How daur ye set your fit upon her -
      Sae fine a lady?
Gae somewhere else and seek your dinner
      On some poor body.
[Robert Burns *To a louse*]

I do not brag that all creation
Is subject to the Flea-ite nation.
I know that parasitic races,
The Ticks and Lucies have their places;
But the imperial race of Flea
Is all surpassing - look at me.
[John Gay *Man and flea*]

One thing is remarkable—that, after some years, the old holes are forsaken and new ones bored; perhaps because the old habitations grow foul and fetid from long use, or because they may so abound with fleas as to become untenantable. This species of swallow, moreover, is strangely annoyed with fleas; and we have seen fleas, bed-fleas (pulex irritans), swarming at the mouths of these holes, like bees on the stools of their hives.
[Gilbert White *The natural history of Selborne*]

When the gnats dance at evening, Scribbling on the air, sparring sparely, Scrambling their crazy lexicon, Shuffling their dumb cabala, Under leaf shadow. Leaves only leaves, Between them and the broad swipes of the sun. Leaves muffling the dusty stabs of the late sun. From their frail eyes and crepuscular temperaments.
[Ted Hughes *Gnat-Psalm*]

Auld Scotland, on thy bonnie face,
Whan Mither Nature gied ye grace,
Lown, birken glens an floery braes,
Wild windy ridges,
To save ye frae deleerit praise,
She gied ye midges.
[W.R. Darling *The pest*]

A 'bug' can mean a number of things (including a flaw in computer software). But its commonest meaning historically has been 'a small insect'. That is the sense in which it is used here. Robert Frost [first extract] introduces another word, when he talks of a 'mite'. This term can be used generically for very small insects, but mites are officially very small arachnids (members, that is, of the same insect group as spiders). Frost is fascinated by the fact that such a minute creature can partake of all the basic animal functions. Mites are often parasitic on animals or plants, and may as part of their activities spread diseases. The same is true of a number of bugs, which is one reason why so many writers have paid attention to them.

The louse is a common example of a bug that is both parasitic and can spread disease. A rubric - the large letters in a manuscript heading - is the most likely place for a louse to get stuck on an old document. However, Steinbeck's identification of the particular kind of louse raises some problems [second extract]. In the first place, the Latin name of a crab louse is actually *Pthirus pubis*. Secondly, as the specific name indicates, crab lice infect the pubic hairs (though they can sometimes be found in head hair or eyelashes). It is the body louse and the head louse that are typically found higher up in the body, and so are more likely to be shed onto documents. It is also the body louse rather than the crab louse that is associated with disease. Steinbeck is using the word 'plague' here rather loosely. The louse is actually associated with epidemics of typhus. It is sometimes difficult to remember how louse-prone our ancestors were. The poor were always affected, but, as Burns [third extract] records, all classes were liable to infestation. The full title of his poem is: 'To A Louse, On Seeing One on a Lady's Bonnet at Church'. Adult lice can be 2mm or more in length, and so are easily visible on an appropriate surface. It was this sighting that led Burns to comment: 'O wad some pow'r the giftie gie us/ To see oursels as ithers see us!'.

Fleas are similar in size to lice, but whereas the latter have specific hosts – thus human lice require human blood - the former may be parasitic on a range of animals [fourth extract]. It is fleas, rather than lice, that are associated with plague. Bacteria causing bubonic plague are transmitted via fleas that bite first rodents and then humans. Fleas are associated with a range of other diseases, as are ticks (a relative of the mite). It is not clear which bug was meant by 'lucies', but they are clearly a related group of blood-sucking parasites. Bird fleas, as Gilbert White notes in his observations of swallows [fifth extract], cluster round the nests of the birds on which they are parasitic. They crawl over the birds in their nests in order to feed, but then leave them, rather than accompanying them on their flights. *Pulex irritans* is as happy to feed on humans as on swallows. It is widely spread both geographically and throughout history. The shortest poem in the English language is said to be by Ogden Nash on the history of fleas: 'Adam/Had 'em'.

Unlike lice and fleas, gnats are flying insects. There are many species, each with its own pattern of behaviour. Some are parasitic on plants and animals. Females of some of the parasitic species feed on blood and so can transmit disease to humans and livestock. Clouds of gnats, typically seen at dusk [sixth extract], are males on mating flights. Gnat eggs are laid in various environments, depending on the species, but shady, damp sites are often popular. Gnats have been a popular topic for poets ever since Virgil. Their swarming has aroused particular comment: Hermann Hesse's *A swarm of gnats*, for example.

'Midge' is a generic term covering various kinds of small fly. As with gnats, different types have different habits. Not all midges bite, but some are blood-suckers, targeting humans and other mammals (and so they can sometimes transmit disease). The best-known of these biting midges in Britain is the highland midge [final extract]. This particular species favours marshy areas, and is common throughout the summer in North-West Scotland (though it can be found elsewhere in Britain, and also across Northern Europe and beyond). Its fame and name derive from its occurrence at the height of the tourist season in one of the most scenic parts of the UK. As with gnats, it is the female midges who do the biting. Unlike gnats, though midges also tend to swarm around dusk, it is the females who congregate. Windy days with low humidity do provide some protection from midges. It has been suggested (tongue-in-cheek) that the 'Highland fling' dance originally derived from the actions of Scottish people as they tried to protect themselves from the attention of midges.

Carbon, in fact, is a singular element: it is the only element that can bind itself in long stable chains without a great expense of energy, and for life on earth (the only one we know so far) precisely long chains are required. Therefore carbon is the key element of living substance: but its promotion, its entry into the living world, is not easy and must follow an obligatory, intricate path, which has been clarified (and not yet definitively)only in recent years. If the elaboration of carbon were not a common daily occurrence, on the scale of billions of tons a week, wherever the green of a leaf appears, it would by full right deserve to be called a miracle. [Primo Levi *The periodic table*]

When sunlight bathes the chloroplast, and photons are absorbed
The energy's transduced so fast that food is quickly stored,
Photosynthetic scenery traps light the spectrum through
Then dark pathway machinery fixes the CO_2 [Harold Baum *The biochemists' songbook*]

Secondly, my Hiawatha
Made with cunning hand a mixture
Of the acid pyrro-gallic,
And of glacial-acetic,
And of alcohol and water
This developed all the picture.
[Lewis Carroll *Hiawatha's photographing*]

Oil and ointment, and wax and wine,
And the lovely colours called aniline:
You can make anything, from a salve to a star
(If you only know how), from black coal tar.
[Punch (1859) *Black coal tar*]

So Kekulé the chemist watched the weird rout
Of eager atom-serpents writhing in and out
And waltzing tail to mouth. In that absurd guise
Appeared benzine and aniline, their drugs and their dyes.
[Robert Graves *The marmosite's miscellany*]

Take carbon for example then
What shapely towers it constructs
To house the hopes of men!
What symbols it creates
For power and beauty in the world
Of patterned ring and hexagon
[A.M. Sullivan *Atomic architecture*]

Jamf, among others, then proposed, logically, dialectically, taking the parental polyamide sections of the new chain, and looping them around into rings too, giant 'heterocyclic' rings, to alternate with the aromatic rings. This principle was easily extended to other precursor molecules. A desired monomer of high molecular weight could be synthesized to order, bent into its heterocyclic ring, clasped, and strung in a chain along with the more 'natural' benzene or aromatic rings. Such chains would be known as 'aromatic heterocyclic polymers.' [Thomas Pynchon *Gravity's rainbow*]

Carbon   DISCUSSION

Primo Levi, himself a chemist, explains succinctly why carbon is such an important element [first extract]. It has the ability - unique amongst elements - of linking to itself more or less *ad infinitem*. In consequence, there are more chemical compounds based on carbon than on all the other elements combined. Carbon forms the basis for all known organic life, which is why the study of its compounds is labelled 'organic chemistry'. Though carbon itself is a solid, some of its compounds are gases. The most notable of these is carbon dioxide, which plants absorb in the process of photosynthesis. This process has been of great importance not only for the evolution of plants, but also for the evolution of animals, since a by-product of photosynthesis is the release of oxygen into the atmosphere. Effectively, carbon dioxide and water combine to produce sugar, which stays in the plant, and oxygen, which is emitted. The process requires the input of energy, which comes from sunlight (with the consequence that photosynthesis is essentially a daytime activity). Primo Levi published *The periodic table* in the 1970s. In the post-war years, a great deal had been discovered about how the photosynthetic process works. As he says, however, much still remained to be discovered.

Photosynthesis takes place at sites in the leaf called chloroplasts. It employs the carbon compound chlorophyll, whose green pigment accounts for the colour of leaves. It was included by Harold Baum in one of his lyrics describing biochemical pathways (in this case, the photosynthesis pathway). (The words are intended to be sung to the tune of *Auld Lang Syne*.) Baum was professor of biochemistry in the University of London. His songbook came out first in the 1980s, and the songs have been sung with pleasure by successive generations of biochemistry students.

Lewis Carroll was a keen photographer, and in this stanza [third extract] he lays out the carbon compounds used to develop his photographs. Pyrro-gallic was derived from phenol (also known as carbolic acid) and was a popular developing agent in the nineteenth century. It fell out of favour in the following century because its effects were sometimes erratic, though a modified form has been used more recently in black and white photography. Acetic acid is the pure form of vinegar ('acetum' being the Latin word for vinegar). Glacial acetic is simply acetic acid with no admixture of water. Finally, alcohol must be the most famous carbon compound of them all.

Coal tar is a by-product in the production of coke and gas, both of which were in increasing demand in the nineteenth century. It is a complex mixture of organic compounds. One of the first of these to be studied was aniline, which became famous for the colourful compounds that it formed. In the mid-1850s, W.H. Perkins in England discovered a mauve compound of aniline which he put into industrial production. This was the first synthetic dye, and gave a much stronger colour than previous natural dyes. As Mr. Punch says, however, this was only one of the potential uses of coal tar [fourth extract]. It should be added that the name 'carbon' actually derives from the Latin for 'coal'.

It was found that the idea of organic molecules as a chain of carbon atoms did not seem to fit all the compounds discovered. The problem was resolved in the 1860s, when the German chemist, August Kekulé, showed that benzene actually consisted of a ring of carbon atoms. It was soon realised that the same was true of other compounds, such as phenol and analine. Many years later, Kekulı explained that the idea had come to him in a day-dream in which he saw a snake turning to bite its own tail - as Robert Graves describes [fifth extract]. Kekulé drew his carbon ring as a hexagon [a six-sided figure] with a carbon atom at each corner. Molecules with this kind of layout are often called 'aromatic' compounds, because some of them have distinctive smells.

Drawings of carbon compounds in books or articles are attempts to represent the structure of the molecules diagramatically. However, such diagrams are two-dimensional, whereas molecules are three-dimensional entities. So structural chemistry nowadays must be concerned with their 3-D geometry, which is what Sullivan is describing in the sixth extract.To understand the chemical work that Pynchon is describing in the final extract requires a look at some of the terms he employs. Carbon's ability to bond to itself is reproduced in some of the molecules that the process produces. A single molecule - a monomer - can bind to others to produce a much larger molecule (therefore called a polymer). In this case, the 'polyamide sections' provide the basis for this growth. A 'heterocyclic ring' differs from an aromatic ring in that not all the atoms forming the ring are carbon atoms. What Pynchon describes is typical of the research effort required in attempting to synthesise new drugs.

Geography is about maps,
But Biography is about chaps.
[Edmund Clerihew Bentley　　*Biography for beginners*]

Whilst my physicians by their love are grown
Cosmographers, and I their map, who lie
Flat on this bed
[John Donne　　*Hymn to God my God, in my Sickness*]

"That's another thing we've learned from your Nation," said Mein Herr, "map-making. But we've carried it much further than you. What do you consider the largest map that would be really useful?" "About six inches to the mile." "Only six inches!" exclaimed Mein Herr. "We very soon got to six yards to the mile. Then we tried a hundred yards to the mile. And then came the grandest idea of all! We actually made a map of the country, on the scale of a mile to the mile!"
[Lewis Carroll　*Sylvie and Bruno concluded*]

Are they assigned, or can the countries pick their colors?
What suits the character or the native waters best.
Topography displays no favorites; North's as near as West.
More delicate than the historians' are the map-makers' colors.
[Elizabeth Bishop　　*The map*]

But Islands of the Blessèd, bless you son, I never came upon a blessèd　one.
[Robert Frost　*A witness tree*]

So geographers, in Afric maps,
With savage pictures fill their gaps,
And o'er unhabitable downs
Place elephants for want of towns.
[Jonathan Swift　*On poetry: a rhapsody*]

I stand here confessed as a contemporary of the Great Lakes of Africa. Yes, I could have heard of their discovery in my cradle, and it was only right that, grown to a boy's estate, I should have in the later sixties done my first bit of map-drawing and paid my first homage to the prestige of their first explorers. It consisted in entering laboriously in pencil the outline of Tanganyika on my beloved old atlas, which, having been published in 1852, knew nothing, of course, of the Great Lakes.
[Joseph Conrad　*Geography and some explorers*]

The figure of the Earth was the topic for after noon.
First my paper; then yours,
With the hills and valleys of the geoid
Mapped as never before.
[Desmond King-Hele　*For Imre Izsak, Earth shaper*]

The doctor opened the seals with great care, and there fell out the map of an island, with latitude and longitude, soundings, names of hills and bays and inlets, and every particular that would be needed to bring a ship to a safe anchorage upon its shores. It was about nine miles long and five across, shaped, you might say, like a fat dragon standing up, and had two fine land-locked harbours, and a hill in the centre part marked "The Spy-glass."
[Robert Stevenson　*Treasure Island*]

Cartography covers the study and practice of making maps, and, as the clerihew tells us, maps are at the heart of geography [first extract]. Indeed, Ptolemy wrote a treatise on cartography in the 2nd century AD which he entitled *Geographia*. John Donne [second extract] refers to his doctors as 'cosmographers', a term used nowadays mainly for those who map the universe; but it was also applied to those who studied a part of the cosmos and, more particularly, the Earth (our concern here). Maps are used for many purposes: to display administrative boundaries, physical features, statistical data, historical changes, and so on. But all maps face a problem. The Earth is a spheroid, whereas we expect the maps representing it to appear on a flat surface. This is the expectation noted by Donne. To go from one to the other requires some method of projection: the various possibilities form a basic element of cartography. One of the most popular is the Mercator projection, originally devised by the Flemish geographer, Gerardus Mercator, in 1569 as an aid to navigation.

One of the basic decisions when making a map is its scale. In Britain, the Ordnance Survey has traditionally made a division between large-scale and small-scale maps. A large-scale map meant six inches to the mile (translated in the 1980s into metric form) or larger – the scale mentioned by Lewis Carroll [third extract]. Obviously, a larger scale means more detail, but it also means a smaller area covered, if the map is to be handled with any ease. Carroll's German professor admits the latter is a problem with the 1:1 scale he describes. Jorge Luis Borges took over the idea a century later, and his narrator also accepts that the result was 'cumbersome'. Another decision relates to the colours typically used to distinguish different features (such as countries) on the map. Some map colours are fairly standard – water is usually shown in various shades of blue, for example, while topography is often shown by yellows, greens and browns. Political maps typically show the greatest range of colours (though British maps usually showed the Commonwealth in pink). Elizabeth Bishop clearly had these points in mind when she wrote *The map*, the first poem in her first collection *North and South* [fourth extract].

Robert Frost's epigram looks back to classical times [fifth extract]. As Ptolemy's map shows, the area around the Mediterranean was quite well-known, but elsewhere often involved a mythological element. The Islands of the Blessed (otherwise known as the Fortunate Isles, or, as in Tennyson's *Ulysses*, the Happy Isles) are an example. This supposed home for heroes lay somewhere in the Western ocean: many places have been suggested as possible sites – Madeira, the Canary Islands, and so on. As travel and exploration increased, so maps of the Earth improved. The main uncertainties tended to be the interiors of continents such as Africa. Maps often contained little pictures – for example, Zephyr puffing out a West wind. As Swift says, these could be used to cover ignorance of what was actually there [sixth extract].

The nineteenth century was a time when the exploration of unknown areas of the Earth became common. Joseph Conrad had, as one of his characters in *Heart of darkness* says, 'a passion for maps'. The expeditions that placed the Great Lakes on maps of Africa took place primarily in the 1850s and 1860s. Conrad's atlas would therefore not show them: he, himself, was born in 1857, so the African expeditions would be recent or current news for him [seventh extract]. Lake Tanganyika was first surveyed by Burton and Speke in 1858.

With the advent of satellites in the twentieth century, global mapping from space became feasible. One of the first results – derived in the 1960s from a study of satellite orbits – was a much improved idea of the Earth's shape. One of the pioneers of such study was the British scientist, Desmond King-Hele. Apart from his work at Farnborough, King-Hele is a leading expert on Erasmus Darwin and also writes poetry. The poem quoted here records the death of an American colleague at a conference they were both attending. Fiction is full of maps. The plot of *Treasure Island* hinges on the discovery of a map [final extract]. In this case, the map was drawn by the author. Similarly, William Faulkner drew his own map of Yoknapatawpha County as a backing for his writings. Other authors employ map-makers. Tolkien and C.S. Lewis, for example, drew their own maps of Middle-Earth and Narnia, but also involved professional illustrators.

The little boys then went out, and returned to a diversion they had been amusing themselves with for several days, the making a prodigious snow-ball. They had begun by making a small globe of snow with their hands. This they turned over and over, till, by continually collecting fresh matter, it grew so large that they were unable to roll it any farther. Here Tommy observed that their labours must end, "for it was impossible to turn it any longer." "No," said Harry, "I know a remedy for that." So he ran and fetched a couple of thick sticks, about five feet long, and giving one of them to Tommy, he took the other himself. He then desired Tommy to put the end of his stick under the mass, while Harry did the same on his side; and then, lifting at the other end, they rolled the heap forward with the greatest ease.
[Thomas Day *The history of Sandford and Merton*]

So Tom went home with Ellie on Sundays, and sometimes on week-days, too; and he is now a great man of science, and can plan railroads, and steam-engines, and electric telegraphs, and rifled guns, and so forth; and knows everything about everything, except why a hen's egg don't turn into a crocodile, and two or three other little things which no one will know till the coming of the Cocqcigrues. And all this from what he learnt when he was a water-baby, underneath the sea.
[Charles Kingsley *The water-babies*]

Blue and yellow smoke shot out from every part of the machine. Wheels whizzed. Levers clicked. Little bits of stuff went buzzing up and down and round and round. And far beneath them the landscape rushed by quicker and quicker until at last they could see nothing but a grey haze all round them. On went the machine, but nothing else happened. On and on they whirled, and nothing happened. And it kept on happening over and over again, till everything was so nothing that neither of them could notice anything.
[Norman Hunter *The incredible adventures of Professor Branestawm*]

Resting the telescope on a rock, he had at last got a steady view. He had seen a black and white dappled back, a greyish head and nape, a neck striped black and white at the sides and a broad stripe of black down the throat. There could be no doubt about it. 'Black-throated diver', he had whispered reverently, put down the telescope, pulled out his pocket-book and added that name to his list of birds seen on the cruise.
[Arthur Ransome *Great Northern?*]

I have serious reason to believe that the planet from which the little prince came is the asteroid known as B-612. This asteroid has only once been seen through the telescope. That was by a Turkish astronomer, in 1909. On making his discovery, the astronomer had presented it to the International Astronomical Congress, in a great demonstration. But he was in Turkish costume, and so nobody would believe what he said.
[Antoine de Saint-Exupéry *The little prince*]

After tea Mr Dobbs asked to be changed into an atom of caesium, because that is the fattest kind of atom, though of course much too small to see. But this was a specially big one, about four feet across. He was quite round, and full of bits buzzing round the middle. They were buzzing so quick that one could never see them properly. There was one bit that sometimes came right out and once or twice shot across the room. Mr Dobbs said that happens when atoms get excited. However, he always got that bit back again, and gave a lovely flash of light when he did.
[J.B.S. Haldane *My friend Mr. Leakey*]

Above them tower the huge atommic PILE which they hav constructed in the MUSICK room. There are 10 bombs - a litle one which trigger off a bigger one, which trigger off the next one until you get to the last which is super coolossal. 'Well?' sa prof. molesworth. 'You seme thortful?' 'It is just that it appere to be going to a lot of trubble just to heat the gym.'
[Geoffrey Willans and Ronald Searle *Whizz for atomms*]

Sandford and Merton was published in three parts during the 1780s. Its author, Thomas Day, was a believer in Rousseau's educational theories, and these are worked out in the story which describes how Sandford, a farmer's boy, seeks to instruct Merton, the spoilt son of a wealthy merchant. Sandford's instruction includes simple applications of science: in the first extract, this is the principle of the lever. Day claimed that he wrote *Sandford and Merton* because so few books were then available to help children advance their reading. The book remained popular throughout the nineteenth century, and was even used as a basic science primer.

The water-babies, which achieved success as soon as it appeared in the 1860s, was partly written as a counterblast to all the factual books that were then being published. Kingsley believed that children also needed fairy stories. Yet his own book contains considerable natural history, and is the first children's book to introduce evolutionary ideas. (Kingsley was a keen marine naturalist, and had previously published a popular book about the seashore.) At the end of the book, the hero, Tom, is resuscitated and becomes, so the second extract says, a 'great man of science'. However, of the four examples noted, only the telegraph involved much scientific input in Kingsley's day. In fact, it is the question Tom cannot answer which is purely scientific. (*Cocqcigrue*, incidentally, was a French word for monster and the phrase means something that was unlikely to occur).

Kingsley like so many others up to the present day, failed to distinguish between 'scientists' and 'inventors'. A more recent example is Professor Branestawm, who first appeared in the 1930s [third extract]. Since Branestawm has no affiliation, 'professor' is simply a title designed to suggest scientific repute. The inventions he makes - in this case, a machine for travelling through time - appear without reference to science: the fun lies in exploring what happens next. But another strand is the interest in natural history. Adventure stories, where children explore the countryside, may well contain insights into natural history. For example, Arthur Ransome's stories often include some ornithology. The fourth extract comes from Ransome's last book. Dick Callum, who features in this extract, is the scientifically inclined member of the group. In this tale, which is concerned with bird protection, he plays a leading role.

Both science fantasy and science fiction - younger Victorian readers were great addicts of Verne stories - continued as genres in the twentieth century. Antoine de Saint-Exupéry was well-known in both French literary and aviation circles in the years leading up to the Second World War. *The little prince* was published in 1943, the year before he died. In the fifth extract, the hero is identified as having come from a small asteroid. The asteroids - large numbers of rocks of various sizes - follow orbits between Mars and Jupiter. The first was discovered in 1801; a century later, so many were known that they had to be assigned numbers. (Though the numbering system agreed internationally was not that used here.) Matters relating to astronomy were discussed at individually arranged International Astronomical Congresses until a permanent body - the International Astronomical Union - was founded in 1919.

For the most part, children's fiction containing science has not been written by scientists. There are exceptions - one being this book by J.B.S. Haldane [sixth extract]. Haldane believed that magicians in the old days were simply trying to do what scientists nowadays actually did do. His tales of the magician, Mr. Leakey, include various asides about science. The one quoted here relates to a party where the physicist, Dobbs, asks Leakey to turn him into a caesium atom. Haldane's description of such an atom actually provides a reasonable sketch of its properties.

After the Second World War, references to science were liable to appear in unexpected places in children's books. The final extract comes from one of the stories about Nigel Molesworth, the archetypal naughty boy of the 1950s. It reflects the common confusion at the time between the controlled power of the atomic pile and the uncontrolled power of the atomic bomb.

A little yellow bird (it is either a species of the alauda trivialis, or rather perhaps of the motacilla trochilus] still continues to make a sibilous shivering noise in the tops of tall woods. The stoparola of Ray (for which we have as yet no name in these parts) is called, in your Zoology, the fly-catcher. There is one circumstance characteristic of this bird, which seems to have escaped observation, and that is, that it takes its stand on the top of some stake or post, from whence it springs forth on its prey, catching a fly in the air, and hardly ever touching the ground, but returning still to the same stand for many times together.
[Gilbert White *The natural history and antiquities of Selborne*]

There thistles stretch their prickly arms afar,
And to the ragged infant threaten war;
There poppies, nodding, mock the hope of toil,
There the blue bugloss paints the sterile soil;
Hardy and high, above the slender sheaf,
The slimy mallow waves her silky leaf;
O'er the young shoot the charlock throws a shade,
And the wild tare clings round the sickly blade;
[George Crabbe *The village*]

Within the reach of every one of you are wonders beyond all poets' dreams. Not a hedge-bank but has its hundred species of plants, each different and each beautiful; and when you tire of them--if you ever can tire--a trip into the meadows by the Thames, with the rich vegetation of their dikes, floating flower-beds of every hue, will bring you as it were into a new world, new forms, new colours, new delight......   In entomology, too, if you have any taste for the beauties of form and colour, any fondness for mechanical and dynamical science, the insects, even to the smallest, will supply endless food for such likings; while their instincts and their transformations.
[Charles Kingsley *Scientific essays and lectures*]

"Do the duty which lies nearest you, and catch the first beetle you come across, is my motto; and I have thriven by it for some hundred years. Now I must go on. Dear me, while I have been talking to you, at least nine new species have escaped me." And on went the giant, behind before, like a bull in a china-shop, till he ran into the steeple of the great idol temple.....But little he cared; for as soon as the ruins of the steeple were well between his legs, he poked and peered among the falling stones, and shifted his spectacles, and pulled out his pocket-magnifier, and cried - "An entirely new Oniscus, and three obscure Podurellae! Besides a moth which M. le Roi des Papillons (though he, like all Frenchmen, is given to hasty inductions) says is confined to the limits of the Glacial Drift. This is most important!"
[Charles Kingsley *The water babies*]

The wood I sought was in deep shelter sunk,
Though clematis leaves shone with a glossy sweat
And creeping over ground and up tree-trunk
The ivy in the sun gleamed bright and wet.

From that small sun patching the wood with light--
O strange to think--hung all things that have breath,
Trees, insects, cows, even moths that fly by night
And man, and life in every form--and death.
[Andrew Young *August*]

Many clergy down the centuries have had an interest in natural history. It has been especially noticeable in the Church of England. One reason may be the way that its church community is organised, with an educated person [= parson] associated with every parish. Traditionally, parsons have remained in their parishes for a number of years. In less hurried times, more especially during the eighteenth and nineteenth centuries, this gave them time to become thoroughly acquainted with the locality in which they lived. Such knowledge might well include a study of its natural history. Clergymen have played an important role in the collection and classification of fauna and flora in England. It is worth remembering that Charles Darwin went to Cambridge in the expectation that he might join the ranks of the clergy.

A few of these clergymen-naturalists also have a place in literature. The archetypal figure is Gilbert White. White was born in Selborne, a village in Hampshire, where he later became the curate-in-charge. He was interested in every aspect of his parish, but especially in the fauna, and he wrote up his observations in letters to friends. These were published in 1798 as *The natural history and antiquities of Selborne*, from which the first quotation comes. The Latin names reflect the fact that White was influenced by the classification system of his contemporary, Linnaeus. (The English names of the birds he mentions are, respectively, tree pipit and willow warbler.) John Ray, in the latter half of the seventeenth century, was one of the earliest parson-naturalists. His work on classification preceded that of Linnaeus: the stoparola, as White says, was Ray's name for the flycatcher. As this extract illustrates, White helped pioneer the studies of animal behaviour that have become so popular in more recent times.

George Crabbe lived in more than one place, but is often associated with Suffolk - perhaps because he is best remembered for the story of Peter Grimes. He was a younger contemporary of Gilbert White; early on, he was a friend of Samuel Johnson, and later he became a friend of Wordsworth. His career as a clergyman was spent mainly in the country, and his poetry reflects the hard lives of ordinary country people. Thus the extract given here shows his knowledge of the local fauna, but he presents it in the context of the weeds that afflict the lives of countryfolk. His reputation as a naturalist actually came mainly from his knowledge of beetles.

The most widely known clergyman-naturalist of the nineteenth century was probably Charles Kingsley. His interests covered all areas of natural history, though he was particularly knowledgeable about marine biology.
His activities as a naturalist were especially prominent in the early 1870s, when he was a canon of Chester Cathedral. While there, he founded the Chester Society for Natural Science, Literature and Art, becoming famous for the field trips he organised, some involving several hundred people. The first extract from his writings reflects his enthusiasm for natural history as a part of education. He, himself, had been inspired by one of his teachers, Charles Johns, who was also a clergyman-naturalist. Johns wrote some of the most popular natural history books of the nineteenth century. Kingsley was aware that a devotion to natural history, particularly the collecting of specimens, had its ridiculous side. This is reflected in the extract from *The water babies*. Here, an otherwise friendly giant is creating havoc by his single-minded devotion to collecting. Thomas Huxley, whose work Kingsley also parodied, was a close friend, and Kingsley, as *The water babies* reveals, was sympathetic to Darwin's ideas on evolution. Darwin was, of course, himself an avid collector of specimens.

In the increasingly frenetic twentieth century, clergymen became less and less able to devote time to other activities such as natural history. One who did was Andrew Young, a noted poet who died in 1971. The final extract comes from one of his nature poems. Young, an amateur botanist, was said to have personally seen more British plants than anyone else in the country. He started life as a Scottish Presbyterian minister, but was later ordained in the Church of England. One of the interesting aspects of the clergymen-naturalists in their heyday was how relatively few came from the Nonconformist churches.

Sunsets are quite old-fashioned. They belong to a time when Turner was the last note in art. Upon the other hand they go on. Yesterday evening Mrs Arundel insisted on my going to the window and looking at the glorious sky.... And what was it? It was simply a very second-rate Turner.
[Oscar Wilde *The decay of lying*]

O painted clouds ! sweet beauties of the sky,
How have I view'd your motion and your rest,
When like fleet hunters ye have left mine eye,
In your thin gauze of woolly-fleecing drest;
Or in your threaten'd thunder's grave black vest
[John Clare *To the clouds*]

But Howard gives us with his clear mind
The gain of lessons new to all mankind;
That which no hand can reach, no hand can clasp
He first has gained, first held with mental grasp.
[Goethe *In honour of Howard*]

We find, however, together with this general delight in breeze and darkness, much attention to the real form of clouds, and careful drawing of effects of mist; so that the appearance of objects, as seen through it, becomes a subject of science with us; and the faithful representation of that appearance is made of primal importance, under the name of aerial perspective. The aspects of sunset and sunrise, with all their attendant phenomena of cloud and mist, are watchfully delineated; and in ordinary daylight landscape, the sky is considered of so much importance, that a principal mass of foliage, or a whole foreground, is unhesitatingly thrown into shade merely to bring out the form of a white cloud. So that, if a general and characteristic name were needed for modern landscape art, none better could be invented than "the service of clouds."
[John Ruskin *Modern painters*]

I wield the flail of the lashing hail,
And whiten the green plains under,
And then again I dissolve it in rain,
And laugh as I pass in thunder.
....
I am the daughter of Earth and Water,
And the nursling of the Sky;
I pass through the pores of the ocean and shores;
I change, but I cannot die.
[Shelley *The cloud*]

Down the blue night the unending columns press
In noiseless tumult, break and wave and flow
[Rupert Brooke *Clouds*]

The patrol leader climbed for some time ..... making his way towards one of the strips of blue sky that here and there showed through the mass of cumulus. They entered the opening at five thousand feet, and then corkscrewed upwards, climbing steeply as though through a hollow tube to the top side of the cloud.
[W.E. Johns *Biggles learns to fly*]

Weather has always been a popular topic for discussion, especially in the British Isles. Clouds figure prominently in such discussions, especially as indicators of likely weather conditions. So it is hardly surprisingly that clouds often feature as a topic in both poetry and prose. Sunlight on clouds at sunrise and sunset attract particular mention because of their vivid colours. The first extract reveals an unimpressed Oscar Wilde (though he and his contemporaries were well acquainted with the old saying: 'Red sky at night; shepherds delight/Red sky in the morning; shepherds warning'). The differing colours and shapes of clouds were, of course, very familiar to people like John Clare [second extract] who spent much of their lives out-of-doors. But their descriptions were frequently vague and unsystematic.

This situation changed at the beginning of the nineteenth century, when Luke Howard - sometimes called the 'father of modern meteorology' - published his paper on *The modification of clouds*. This classified the clouds into different groups and gave them the names which are, for the most part, still in use today (such as cirrus and cumulus). The paper was subsequently translated into German and published in the *Annalen der Physik* in 1815. There it came to the attention of Goethe, who already had an interest in clouds and soon waxed enthusiastic about the new classification. In 1820 he published a poem in honour of Howard, a translation of a part of which forms the third extract.

Howard's essay also attracted the attention of artists both in Britain and abroad. The most important of these was John Constable. Howard's classification of clouds fitted in with Constable's own way of thinking. He claimed: 'Painting is a science, and should be pursued as an enquiry into the laws of nature'. Apart from the depiction of clouds in his landscape paintings, he also became famous for his 'skyscapes' of the different types of cloud. Ruskin, writing some years after Constable's death, strongly stressed the importance of clouds for landscape painters [fourth extract]. (Ruskin's prime allegiance was to Turner, whose concern was more with light effects, than with types of cloud. This is, perhaps, why Wilde thought a real sunset compared poorly with a Turner painting.)

In terms of literature, Shelley was the British writer most influenced by Howard. Shelley's poem *The cloud* consists of a series of stanzas which can be related to Howard's descriptions of clouds. Thus the four lines quoted first from the poem [fifth extract] refer to nimbus clouds (defined as dark grey clouds that give precipitation of some kind). The second four lines describe the water cycle - the way that water is precipitated from the atmosphere onto the land and sea and then evaporates back again. The existence of this cycle had been recognised from antiquity, but many of the details were still vague in the early nineteenth century. For example, there was no good explanation for the formation of raindrops in the atmosphere. One theory - originally suggested by Benjamin Franklin - involved the electricity of the atmosphere. This would have pleased Shelley who had been fascinated by electricity from his schooldays.

Shapes and colours are not the only characteristics of clouds. Their positions can also sometimes be significant. Thus the 'unending columns' of Rupert Brookes' poem are probably 'cloud streets'. These are long, parallel lines of cumulus clouds, oriented almost along the direction of the prevailing wind. They are the result of long rolls of convection in the atmosphere which can occur when conditions are right. Clouds can also be distinguished by their height in the atmosphere. Cumulus clouds and cloud streets form at low levels - up to 2 Km. Cirrus clouds, by way of contrast, can be found up to 13 Km. All the descriptions so far have been of clouds seen from the ground. Though it has always been possible to look down on low-lying clouds from a mountain top, scanning clouds from above became a much commoner activity in the nineteenth century as high-altitude ballooning became feasible. The advent of powered flight at the beginning of the twentieth century meant that clouds could even be explored – as in the First World War yarn that provides the last extract.

When beggars die there are no comets seen;
The heavens themselves blaze forth the death of princes.
[William Shakespeare *Julius Caesar*]

Incens'd with indignation Satan stood
Unterrify'd, and like a comet burn'd
That fires the length of Ophiuchus huge
In th' arctic sky, and from his horrid hair
Shakes pestilence and war.
[John Milton *Paradise lost*]

And though men read no gospels in these signes,
Yet all professions are become divines ;
All weapons from the bodkin to the pike,
The masons rule and taylors yard alike
Take altitudes
[Richard Corbet *A letter sent from Dr. Corbet to Sir Thomas Ailesbury*]

He, first of men, with awful wing pursued
The comet through the long elliptic curve,
As around innumerous worlds he wound his way
[James Thomson *To the memory of Sir Isaac Newton*]

Could he, whose rules the rapid comet bind,
Describe or fix one movement of his mind!
Who saw its fires here rise and there descend,
Explain his own beginning or his end?
[Alexander Pope *Essay on man*]

I came in with Halley's comet in 1835. It is coming again next year, and I expect to go out with it. It will be
the greatest disappointment of my life if I don't go out with Halley's comet. The Almighty has said, no doubt:
'Now here are these two unaccountable freaks; they came in together, they must go out together
[A.B. Paine *Mark Twain*]

Their Apprehensions arise from several Changes they dread in the Celestial Bodies. For Instance; ... That,
the Earth very narrowly escaped a Brush from the Tail of the last Comet, which would have infallibly
reduced it to Ashes; and that the next, which they have calculated for One and Thirty Years hence, will
probably destroy us.
[Jonathan Swift *Gulliver's travels: Voyage to Laputa*]

I had to look, and then I had to listen, how that this scarce-visible intruder was to be, was presently to be,
one of the largest comets this world has ever seen, how that its course must bring it within at most - so many
score of millions of miles from the earth, a mere step, Parload seemed to think that; how that the
spectroscope was already sounding its chemical secrets, perplexed by the unprecedented band in the green,
how it was even now being photographed in the very act of unwinding--in an unusual direction--a sunward
tail (which presently it wound up again).
[H.G. Wells *In the days of the comet*]

In astrological terms, any unexpected change in the heavens could be significant. The appearance of a bright comet was especially important. Though comets might occasionally have a benign influence, most were regarded with dread. In particular, they were thought to warn of a coming change of governance, often via the death of a leader. So Caesar's comet - one of the brightest comets reported in antiquity - is interpreted by Caesar's wife as a warning to her husband in Shakespeare's *Julius Caesar* [first extract]. (The comet actually appeared after his death.) Shakespeare makes a similar reference to the significance of comets in *Henry VI Part I*. It made sense therefore for Milton in *Paradise Lost* [second extract] to compare Satan with a comet - huge, evil and attempting to change the governance of the world. However, Milton seems to be a little mixed up in his astronomy. Ophiucus (the Serpent Bearer) is not a northerly constellation: it actually crosses the celestial equator. It is usually pictured as a man holding a snake (the neighbouring constellation of Serpens). The confusing factor may be that there is another serpent in the heavens - Draco - which does encircle the north celestial pole. In any case, the positioning of the comet is obviously meant to remind readers of Satan's role as a serpent in the Garden of Eden.

The classical picture of the world supposed comets to be phenomena in the upper atmosphere of the Earth. This was cast in doubt when astronomers across Europe measured the position of a bright comet that appeared in 1577. The Danish astronomer, Tycho Brahe, collected the data and found that the comet's path lay further away than the Moon's orbit. His finding was one of the first indications that the traditional picture was wrong. Measuring the position of comets thus became a significant activity. Three comets were seen in 1618: the one that attracted most attention appeared towards the end of the year. It was widely observed and commented on (not least by King James in Britain) though perhaps not as extensively as the poem quoted in the third extract suggests. ('Taking altitudes' was equivalent to saying - measuring its position.) The first line indicates increasing doubts regarding astrological interpretations of comets.

By the end of the seventeenth century, Newton had provided an explanation of how the planets and their satellites moved. To his contemporaries, however, the really exciting thing was that he also explained the apparently arbitrary motions of comets. James Thomson's eulogy of Newton [fourth extract] lists this as its first point. Pope [fifth extract] disliked the contemporary glorification of Newton, but agreed that Newton's explanation of cometary orbits was his most exciting result. Thomson's poem was written immediately after Newton's death in 1727: Pope's appeared a few years later.

The nineteenth century saw an upsurge of interest in the physical properties of comets. It had been known since the sixteenth century that comet tails pointed away from the Sun. It was now realised that the tails consisted of material blown back from the central nucleus of the comet. Should the Earth subsequently pass through the comet's path, this left-behind material would burn up in the atmosphere producing a shower of meteors.

The discovery that Biela's comet crossed the Earth's orbit caused some popular concern. This was nothing new: Swift's Laputans already dreaded the possibility of a collision with a comet's tail in the eighteenth century [seventh extract]. Even in the early twentieth century, there was popular concern when it was reported that the Earth was due to pass through the tail of Comet Halley in 1910. Wells' novel [eighth extract] was published four years prior to this event. He typically turns the predictions upside down, making the result of the collision beneficial to the inhabitants of the Earth. Wells correctly identifies study of cometary spectra as a major nineteenth-century advance in understanding the nature and composition of comets. Comets' tails point away from the Sun; but, occasionally, the relative positions of Earth, comet and Sun give the appearance that the tail is pointing towards the Sun, rather than away.

On two occasions I have been asked, - 'Pray, Mr. Babbage, if you put into the machine wrong figures, will the right answers come out? In one case a member of the Upper, and in the other a member of the Lower, House put this question. I am not able rightly to apprehend the kind of confusion of ideas that could provoke such a question.
[Charles Babbage *Passages from the life of a philosopher*]

The first Professor I saw was in a very large Room, with forty Pupils about him. After Salutation, observing me to look earnestly upon a Frame, which took up the greatest part of both the Length and Breadth of the Room, he said perhaps I might wonder to see him employed in a Project for improving speculative Knowledge by practical and mechanical Operations..... Every one knew how laborious the usual Method is of attaining to Arts and Sciences; whereas by his Contrivance, the most ignorant Person at a reasonable Charge, and with a little bodily Labour, may write Books in Philosophy, Poetry, Politicks, Law, Mathematicks and Theology
[Jonathan Swift *Gulliver's travels*]

UHL was Unit Headline Language, and it consisted of a comprehensive lexicon of all the multi-purpose monosyllables used by headline-writers. Goldwasser's insight had been to see that if the grammar of "ban", "dash", "fear", and the rest was ambiguous they could be used in almost any order to make a sentence, and that if they could be used in almost any order to make a sentence they could be easily randomised. Here then was one easy way in which a computer could find material for an automated newspaper - put together a headline in basic UHL first and then fit the story to it.
[Michael Frayn *The tin men*]

If this assembly of buildings looked like the worst of future college campuses, all-but-treeless, charmless, institutional, aseptic, milk-of-magnesia white, and composed of many windowless buildings and laboratories which seemed to house computers, and did! why the error was in fact natural.
[Norman Mailer *Fire on the Moon* in *Life* magazine]

A man, white-coated, comes to switch me off.
'Something is wrong with our expensive brain.'
Poor pricked balloon! Yes, something has gone wrong:
Smear your white coat with Socrates and Christ!
Yes, switch me off for fear I should explode:
Yes, switch me off for fear yes switch me off
for fear yes switch me off for fear yes switch (finis)
[John Wain *Poem feigned to have been written by an electronic brain*]

Even now, he could not fully accept the idea that Frank had been deliberately killed - it was so utterly irrational. It was beyond all reason that Hal, who had performed flawlessly for so long, should suddenly turn assassin.
[A.C. Clarke *2001*]

But it was fully fifteen seconds before the round plate that she held in her hands began to glow. A faint blue light shot across it, darkening to purple, and presently she could see the image of her son, who lived on the other side of the earth, and he could see her.
[E.M. Forster *The machine stops*]

He also had a device that looked rather like a largish electronic calculator. This had about a hundred tiny flat press buttons and a screen about four inches square on which any one of a million 'pages' could be summoned.
[Douglas Adams *The Hitch-Hikers Guide to the Galaxy*]

Mechanical devices to aid calculation go back a long way, but devices sufficiently complex to be called 'computers' are a more recent invention. In the first half of the nineteenth century - when Babbage was hard at work - a computer usually meant a lad who spent all his time on arithmetical calculations (as at the Royal Observatory Greenwich, for example). Babbage called the calculating device he developed an 'engine' - a complex machine of interlocking gears. Continuing attempts to improve it meant that a fully operational machine was never completed (though working examples have been built since). But Babbage lay down the requirements for an automatic computer, including the idea of a program (on which Ada Lovelace, Byron's daughter, did notable work). As the first excerpt indicates, though Babbage received state support, he had a limited enthusiasm for the legislators.

The idea of a computer was in circulation before the nineteenth century. Swift [second extract] envisaged a machine that could automatically produce texts. (Babbage was concerned with handling numbers, but a computer can equally be used to handle letters.) The ultimate form of what he is outlining here is known as the 'monkey typewriter problem'. The idea is that if a very large number of monkeys were allowed to bash away at random on an equally large number of typewriters, they would, given sufficient time, produce all known texts. (This proposal has generated its own set of literary references over the past century.)

The tin men [third extract] envisages something similar, but in a more restricted area. It is based on the premise that popular newspapers often employ headlines consisting of a series of monosyllables, which sometimes seem only distantly related to what follows. Electronic computers were developed during the Second World War. By 1965, when this novel appeared, they were coming into use to handle a wide variety of activities, both numerical and textual. Frayn's characters work at the William Morris Institute for Automation Research. The name is part of the satire, since Morris actually detested automation. In a similar way, many post-war writers regarded computers with a degree of suspicion. For one thing, computers were seen as having a dehumanising influence. Mailer's description [fourth extract] of the Manned Spacecraft Center in the USA reflects this. (Mailer, like Frayn, was writing in the 1960s.)

One of the recurring themes in discussions of computers is how a really advanced electronic computer - an electronic brain - would compare with a human brain. One possible answer is that a logical machine would have problems coping with the full complexity of a situation as a human does. John Wain's poem, published in the 1950s, envisages a computer in this situation going mad [fifth extract]. A.C. Clarke's novel *2001* [sixth extract], published a decade late, envisages a more intricate response. The space station at the centre of the story is maintained by a computer called HAL (This is said to stand for 'Heuristically programmed Algorithmic Computer'. It was pointed out that advancing each of the letters by one leads to the new sequence 'IBM', but Clarke denied that this was intended.) The computer decides that the whole operation will go more efficiently if the human beings aboard are eliminated. The hero has difficulty in accepting that a computer that has been programmed to help people might then be led by logic to kill them.

Forster's short story [sixth extract] appeared first in 1909. It envisages a world run by a global machine. This machine allows people to interact via a network of videophones. The result is that people stay more and more at home: physical contact decreases and dependence on the machine increases. Eventually, the machine begins to deteriorate and finally stops, bringing chaos. Forster was reacting to what he saw as H.G. Wells' over-optimism in regard to new technology. However, the worries he had about 'the Machine', have been echoed by comments in more recent years about the Internet. Handling graphics via computer-based networks is now commonplace. Because graphics contain a large amount of information, the handling requires powerful computers. Perhaps the most obvious feature of the computer revolution has been the way in which computer power has soared while the space the computer occupies has plummeted. The cartoon series *Dick Tracy* introduced a two-way wrist radio in 1946, upgrading this to a two-way wrist TV in 1964. Both were, of course, well ahead of their time (and are still a matter for experiment today). But the trend to miniaturisation has long been accepted. *The Hitch Hiker's Guide to the Galaxy* [final extract] appeared at the end of the 1970s. It cheerfully foretold the ultimate in miniature multi-purpose computers.

It was an account of an annual gathering of the Geographical Society, and Lord Hollingford had read a letter he had received from Roger Hamley, dated from Arracuoba, a district in Africa, hitherto unvisited by any intelligent European traveller; and about which, Mr. Hamley sent many curious particulars. The reading of this letter had been received with the greatest interest, and several subsequent speakers had paid the writer very high compliments.
[Elizabeth Gaskell *Wives and daughters*]

Of course it was very much to our credit, really, to own such a grandfather; but one mustn't be proud, or show off about it; so we blushed and were embarrassed and changed the subject. It was probably the same wish not to seem presumptuous, which gave my uncles the odd habit of never claiming him as their own father, in conversation with each other. They always said: '*Your* father said so-and-so'; to which the other uncle often answered: 'Well he was *your* father, too.'
[Gwen Raverat *Period piece*]

With Etty, he's bred and dissected
a thousand pigeons, to demonstrate
They're all descended from the same rock dove.
[Ruth Padel *The survival of the fittest*]

7 April 1852
Went to the Zoo.
I said to Him—
Something about that Chimpanzee over there reminds me of you.
[Carol Ann Duffy *Mrs Darwin*]

At the back of the auditorium were curtains, giving upon a museum devoted to the invertebrata. I was told that while Huxley lectured Charles Darwin had been wont at times to come through those very curtains from the gallery behind and sit and listen until his friend and ally had done.
[H.G. Wells *Experiment in autobiography*]

I regret that reviewers have in some cases been inclined to treat the chapters on Machines as an attempt to reduce Mr. Darwin's theory to an absurdity.   Nothing could be further from my intention, and few things would be more distasteful to me than any attempt to laugh at Mr. Darwin.
[Samuel Butler *Erewhon* (Preface to 2nd Ed.)]

What Newton's might could make not clear
Hath Darwin's might not made?
[Algernon Swinburne *The commonweal*]

The PR men of bestiaries
eulogized for centuries
this busy little paragon,
nature's proletarian—
but look here, Darwin said: some ants
make slaves of smaller ants, and end
exploiting in their peonages
the sweating brows of their tiny drudges
[Philip Appleman *The ant*]

Mrs. Gaskell was a distant cousin of Charles Darwin, and her circle of friends included both the Darwin and the Wedgwood families. *Wives and daughters* was left unfinished when she died in 1865 [first extract]. The hero of the novel, Roger Hamley, was modelled on Charles Darwin, being, like him, both a traveller and an avid collector of insects. Darwin's *Origin of species* had appeared in 1859, giving rise to enormous discussion. Unitarians, including Mrs. Gaskell, were more receptive than most religious groups to its arguments. Darwin became a Fellow of the Geographical Society [it added 'Royal' to its title in 1859] on his return from his voyage on the *Beagle* in the 1830s. By the time of Mrs. Gaskell's last novel, the Society was becoming noted for its support of expeditions to Africa.

Gwen Raverat was a granddaughter of Charles Darwin. Towards the end of her life, she published *Period piece* containing reminiscences of her childhood. (She was a leading wood engraver and illustrated the book herself.) Three of Darwin's sons became Fellows of the Royal Society. One of them was George Darwin, Gwen Raverat's father, who became Plumian Professor of Astronomy and Experimental Philosophy at Cambridge shortly before she was born. Gwen says that her own image of her grandfather was a cross between God and Father Christmas. It is evident from her book that all the Darwin family stood in some of awe of Charles [second extract].

Ruth Padel is a great-great-granddaughter of Charles Darwin. Her interest in both science and poetry has led her to write a series of poems on Darwin - effectively a verse biography - based on his own writings. The third extract mentions Darwin's daughter, Henrietta [Etty], who helped him in some of his work. Darwin was fascinated by pigeons and spent some time breeding them. What interested him was that birds of vastly different appearance had all been bred from a single ancestral species. He hoped that his study of their artificial selection would provide insight into the process of natural selection. When he had finished his work, he gave the pigeons to the Natural History Museum in London, where they are still stored.

Darwin's career was founded on his voyage round the world aboard the *Beagle* in the 1830s. At the end of that decade he married his cousin, Emma Wedgwood. They had ten children; perhaps equally importantly, she cared for him during his frequent bouts of illness. Duffy's poem [fourth extract] is one of a series she has written from the point of view of the wives of famous men. It combines in a few words a picture of Darwin, evolution and religion. However, it seems to have little direct connection to Emma. Nor does the date figure as important in the lives either of Emma or Charles (though the 'two' obviously rhymes with the rest).

As a result of his continuing illness, Darwin spent much of his time at home in Kent. He still visited London from time to time, where he met up with old friends, not least T.H. Huxley. Wells [fifth extract] attended Huxley's lectures (at what is now Imperial College) not long after Darwin's death. As his autobiography illustrates, Darwin was a figure of awe to students, more even than to his family. Inevitably, evolutionary ideas were applied to new areas, giving rise particularly to various types of 'Social Darwinism'. In *Erewhon* [sixth extract], Butler applied the concept to machines, suggesting that they, too, might eventually self-evolve. The possibility has become widely discussed in recent years, but (as the extract shows) Butler, at the time, had to reassure Darwin that it was not intended as a parody of his work.

Darwin, unlike Huxley, did not see much point in attacking religion. But several contemporaries outside the world of science seized on this aspect of his work. Swinburne was a leading proponent of atheism. As the seventh extract succinctly illustrates, he believed that Newton had opened up the path to atheism by mechanising the Universe. Darwin, he thought, had completed the work by mechanising life. Appleman [eighth extract] would certainly sympathise with Swinburne, and his fascination with Darwin has led to a number of poems. The one cited here compares the traditional moral picture of the hard-working ant (for example, in *Proverbs* **6**:6) with the amorality suggested by Darwinian evolution.

Yet is my cheek with rosy blushes warm;
Yet are my eyes with sparkling lustre fill'd;
Yet is my mouth replete with murmuring sound;
Yet are my limbs with inward transport fill'd;
And clad with new-born mightiness around.
[Humphry Davy *On breathing nitrous oxide*]

It is said that every excitation is followed by a commensurate exhaustion. That is not so. The excitation caused by inhaling nitrous oxide is an exception at least; it leaves no exhaustion on the bursting of the bubble. The operation of this gas is to prevent the decarbonating of the blood; and, consequently, if taken excessively, it would produce apoplexy. The blood becomes black as ink. The voluptuous sensation attending the inhalation is produced by the compression and resistance.
[S.T. Coleridge *Table talk*]

Or call up sages whose capacious mind
Left in its course a track of light behind;
Point where mute crowds on Davy's lips reposed,
And Nature's coyest secrets were disclosed
[Anna Laetitia Barbould *Eighteen hundred and eleven*]

I beg my kindest compliments to Sir Humphry, and tell him Ill Luck, that direful chemist, never put into his crucible a more indissoluble piece of stuff than your affectionate cousin and sincere well-wisher,
[Walter Scott *Letter to Lady Davy*]

Sir Humphrey Davy
Abominated gravy.
He lived in the odium
Of having discovered sodium.
[Edmund Clerihew Bentley]

"Sir Humphry Davy?" said Mr. Brooke, over the soup, in his easy smiling way, taking up Sir James Chettam's remark that he was studying Davy's Agricultural Chemistry. "Well, now, Sir Humphry Davy; I dined with him years ago at Cartwright's, and Wordsworth was there too -- the poet Wordsworth, you know. Now there was something singular. I was at Cambridge when Wordsworth was there, and I never met him -- and I dined with him twenty years afterwards at Cartwright's. There's an oddity in things, now. But Davy was there: he was a poet too. Or, as I may say, Wordsworth was poet one, and Davy was poet two. That was true in every sense, you know."
[George Eliot *Middlemarch*]

By reference to the 'Diary of Sir Humphrey Davy', it will be seen that this illustrious chemist had not only conceived the idea now in question, but had actually made no inconsiderable progress, experimentally, in the very identical analysis now so triumphantly brought to an issue by Von Kempelen, who although he makes not the slightest allusion to it, is, without doubt (I say it unhesitatingly, and can prove it, if required), indebted to the 'Diary' for at least the first hint of his own undertaking.
[Edgar Allan Poe *Von Kempelen and his discovery*]

Humphry Davy was born in Cornwall in the latter part of the eighteenth century. He started on a medical apprenticeship there, but his growing interest in chemistry led to an appointment at an institute run by Thomas Beddoes in Bristol. The central interest there was the medical uses of gases. Davy's main contribution was to prepare the gas, nitrous oxide, and to study its properties. His work led to its subsequent widespread popularity under the name of 'laughing gas'. Davy was also a devoted writer of poetry. Through Beddoes, Davy met both Coleridge and Southey, who were then living in the area. He and Coleridge became particularly close friends, and Davy came to be seen as another of the Romantic poets. The first quotation is part of a poem by Davy full of enthusiasm for nitrous oxide. (For a time, he became addicted to breathing the gas.) Coleridge, already an opium addict, also found pleasure in breathing nitrous oxide, as did Southey. Both poets contributed brief descriptions of their experiences to Davy's *Researches chemical and philosophical; chiefly concerning nitrous oxide*. Coleridge's description forms the second extract quoted here.

Coleridge introduced Davy to Wordsworth, and they involved Davy in the editing of their *Lyrical Ballads*. Wordsworth's preface to the second edition of these *Ballads* [second extract] was influenced by a lecture that Davy gave on science. For Davy's new fame led to an appointment at the Royal Institution in London, where he rapidly established himself as an outstanding lecturer. The third extract reflects the enthusiasm his lectures inspired. Laetitia Barbould was well placed to assess this. She knew the London scene well, and was acquainted with both Coleridge and Wordsworth. (She subsequently parted company with both. The poem quoted here was actually written in opposition to the Napoleonic wars. It received so much adverse comment that she ceased writing poetry.) In 1812, Davy was knighted and married a wealthy Scottish widow, Jane Apreece. She was a celebrated hostess in Edinburgh, and was, as the fourth extract indicates, a relative of Sir Walter Scott's. Not only was the marriage was a failure; she also battled with Davy's assistant and successor, Michael Faraday.

After nitrous oxide, Davy turned his attention to electrolysis - the passage of electricity through fluids. This led him to the discovery in 1807 of two new elements - potassium and sodium - which brought him renewed fame. E.C. Bentley was the inventor of the clerihew (a short humorous verse form). It is said that the one quoted here was actually the first he wrote. He allegedly composed it during a chemistry class when he was a schoolboy towards the end of the nineteenth century. (He is not the only one to have experienced difficulties in spelling Davy's forename - see the final extract.)

The Royal Institution had actually been set up to examine possible applications of science. Davy was well aware of this, and one of the topics he examined was the application of chemistry to agriculture. Since the latter part of the eighteenth century, there had been a growing concern with the need to improve agricultural practices. In 1812, Davy was asked by the then president of the Board of Agriculture to give a course of lectures on agricultural chemistry. The publication resulting from these was widely read, as George Eliot notes in the penultimate extract. (Both she and her consort, G.H. Lewes, were interested in science: she attended Faraday's lectures at the Royal Institution.) Though Davy came to be regarded as an authority on agricultural chemistry both in Britain and in the United States, the real breakthrough in terms of practical application came with the German chemist, Liebig, after 1840.

Wolfgang von Kempelen was the inventor of a supposed mechanical automaton - a figure in Turkish dress - which could play chess. (In fact, the Turk, as it was later called, concealed a human chess player.) After von Kempelen's death in 1804, the Turk was sold and taken on tours of America during the 1820s and 1830s: Poe saw it when it was exhibited in Virginia. The spoof tale by Poe - from which the final extract comes - imagines that the invention of the automaton was influenced by Davy's discovery of the effects of nitrous oxide on the human body. The Turk was ultimately bought by Dr. Mitchell, Poe's physician, who formed a restoration club to get it back into working order again. This was in 1840, before Poe wrote the story quoted here. By an odd coincidence, Poe's cousin, George Poe, a leading US chemist, was the first person to liquefy nitrous oxide.

Where shall we our great Professor inter,
That in peace may rest his bones?
If we hew him a rocky sepulchre,
He'll rise and break the stones,
And examine each stratum that lies around -
For he is quite in his element underground.
[Richard Whatley *Elegy intended for Professor Buckland*]

London. Michaelmas term lately over, and the Lord Chancellor sitting in Lincoln's Inn Hall. Implacable November weather. As much mud in the streets as if the waters had but newly retired from the face of the earth, and it would not be wonderful to meet a Megalosaurus, forty feet long or so, waddling like an elephantine lizard up Holborn Hill.
[Charles Dickens *Bleak House*]

I've often wanted to explore
The graveyard of the Dinosaur,
And when the British Museum said
They'd like to own a mighty head,
The largest of the saurians dread,
Triceratops (Marsh gave the name),
To Converse County then I came.
[Charles H. Sternberg *In the Laramie*]

Thou shalt see, in his Jurassic tomb,
The Plesiosaurus embalmed;
In his Oolitic prime and his bloom,
Iguanodon safe and unharmed!
[F. Bret Harte *A geological madrigal*]

Reconstructed by Darwin or Owen,
We dwell in sweet Bloomsbury's halls......
Though so cleverly people restore us,
We are bound to confess with a sigh
That the brain of the Ichthyosaurus
Was *never* so good as his eye!
[May Kendall *Ballad of the Ichthyosaurus*]

"This is an excellent monograph by my gifted friend, Ray Lankester!" said he. "There is an illustration here which would interest you. Ah, yes, here it is! The inscription beneath it runs: 'Probable appearance in life of the Jurassic Dinosaur Stegosaurus. The hind leg alone is twice as tall as a full-grown man.' Well, what do you make of that?"
[A. Conan Doyle *The lost world*]

I have never listened to anyone who criticized my taste in space travel, sideshows, or gorillas. When this occurs, I pack up my dinosaurs and leave the room.
[Ray Bradbury *Zen in the art of writing*]

Though large fossil bones had been found long before, recognition of dinosaurs as a distinct group of extinct animals awaited the nineteenth century. William Buckland at Oxford University acquired some of these large bones in the early nineteenth century. After a discussion with the French comparative anatomist, Cuvier, he published a description of them, naming the animal from which they came *Megalosaurus* [= great lizard]. A contemporary subsequently gave it its currently accepted name - *Megalosaurus bucklandii.* Buckland, a clergyman (he later became Dean of Westminster Abbey) as well as a leading geologist, was also a noted eccentric. His ambition was to eat his way through all the animal kingdom (he claimed that moles and bluebottles tasted worst). His visitors were also expected to partake: John Ruskin regretted having missed the toasted mice. He was a charismatic lecturer - one of his listeners wrote the first extract quoted here at about the time that Buckland was working on Megalosaurus. The giant lizard aroused much popular interest. Dickens' novels are not always noted for their scientific content, but, as the second extract shows, he expected his readers to know about Megalosaurus.

Sternberg [third extract] was a noted American fossil collector in the years before and after 1900. He worked initially for Edward Cope, one of the leading American collectors in the latter part of the nineteenth century. Cope's great rival as a fossil collector was Othniel Marsh. Between the 1870s and the 1890s, these two financed various expeditions within the USA to look for fossils, especially dinosaur bones. Their competition, which came to be known as the 'Bone Wars', ruined both of them financially, but it led to the discovery of scores of new dinosaur species, and sparked much public interest. Sternberg's poem was written after the 'Wars' were over. Laramie in Wyoming was one of the fossil dinosaur sites and Sternberg was going there on behalf of the British Museum (Natural History). Triceratops had distinctive horns on its head, which have made it one of the most illustrated dinosaurs. 'Dread' is rather harsh - it had a vegetarian diet.

Dinosaurs were a dominant life form in the Jurassic [208-144 million years ago] and Cretaceous [144-65 million years ago] periods (whence came the title *Jurassic Park* for the book and film). The plesiosaurus was a marine reptile whose bones have been found in Jurassic rocks in both the UK and the USA. The Iguanodon was the second dinosaur to be identified after Megalosaurus. It, too, was a herbivore, and lived around the Jurassic/Cretaceous divide. Bret Harte's specimen was cased in an oolitic rock - most probably limestone. He himself knew something of geology, having been involved in the California gold rush.

May Kendall's best known poem is probably *Lay of the Trilobite,* a satirical look at Darwinian evolution. The poem quoted here also refers to Darwin, though he, in fact, was never greatly concerned with dinosaurs. Richard Owen, on the contrary, was: he suggested the name 'dinosaur'[= terrible lizard] and was an acknowledged expert on the ichthyosaurus [= fish lizard] (which, despite its name, was subsequently realised not to be a dinosaur). The first complete ichthyosaur fossil had actually been found as early as 1811 in Lyme Regis (along what is now labelled the 'Jurassic Coast'). The reference to 'Bloomsbury' also involves Owen. Initially, natural history specimens, including fossil bones, were handled by the British Museum in Bloomsbury. In 1881, under the supervision of Owen, a separate British Museum (Natural History) opened its doors in South Kensington.

After Sherlock Holmes, Professor Challenger is probably Conan Doyle's best-known creation. In this extract from The lost world, he is trying to convince a journalist that dinosaurs are still alive in a remote part of South America. The dinosaur mentioned, Stegosaurus, is actually one of the easiest to identify from the large plates sticking out from its back. It was first recognised by Marsh during the 'Bone Wars', and is yet another large herbivore. The reference to Ray Lankester is interesting. It has been suggested that Lankester - a leading British zoologist in the decades around 1900 - was a model for Challenger (though Conan Doyle, himself, said that his character was based on a physiology professor at the University of Edinburgh). Dinosaurs have been a continuing stand-by for science fiction writers. Ray Bradbury is one of many who have written about them.

Everything here is alive thanks to the living of everything else.
[Lewis Thomas *The medusa and the snail*]

God blesses him, he
Gives praise with his toil,
Lends comfort to me,
And aerates the soil.
[John Updike *Earthworm*]

There is a tree native to Turkestan,
Or further east towards the Tree of Heaven,
Whose hard cold cones, not being wards to time,
Will leave their mother only for good cause;
Will ripen only in a forest fire
[William Empson *Note on local flora*]

The monarchs of the Irish bogs
Succumbed to neither men nor dogs
But (most ecologists agree)
To calcium deficiency.
[R.A. Lewin *Elks, whelks, and their ilk*]

Watching one of these busy companies of small birds at work one is amazed at the thought of the abundance of larval insect life in these oak woods. The caterpillars must be devoured in tens of thousands every day for some weeks, yet when the time comes one is amazed again at the numbers that have survived to know a winged life. On July evenings with the low sun shining on the green oaks at this place I have seen the trees covered as with a pale silvery mist a mist composed of myriads of small white and pale-grey moths.
[W.H. Hudson *Hampshire days*]

It's how you look at population changes in biology. Goldfish in a pond, say. This year there are x goldfish. Next year there'll be y goldfish. Some get born, some get eaten by herons, whatever. Nature manipulates the x and turns it into y. Then y goldfish is your starting population for the following year.
[Tom Stoppard *Arcadia*]

There was a strange stillness. The birds, for example - where had they gone? ….. The roadsides, once so attractive, were now lined with browned and withered vegetation as though swept by fire. These, too, were silent, deserted by all living things. Even the streams were now lifeless.
[Rachel Carson *Silent spring*]

I want the field not to have to prove anything
by statistics of wheatweight.
I want the field to have its own quota
of roe deer, walkers, horses, flies, vetch
[Rose Flint *The field*]

Ecology studies the way that organisms interact both with their environment and with each other. The first extract emphasizes the interdependence of all organisms on each other - a holistic view of the world. Thomas was a senior figure in twentieth-century American medicine, being dean of medicine at both Yale and New York University. He was also a noted essayist who wrote on a variety of topics in biomedicine. The word 'ecology' actually appeared in Charles Darwin's lifetime. It was coined by one of his supporters not long after the publication of *On the origin of species*. Darwin saw competition as the main motivator of evolution, but early ecologists were more interested in the question of interdependence. However, Darwin, though he never seems to have used the term *ecology*, did work in this area: not least in his final major project in which he studied the interaction of earthworms with their environment. Part of this interaction is reflected in John Updike's poem [second extract].

Ecology, being a wide-ranging topic, includes many specialist areas within its scope. One such is 'fire ecology'. Though fires are obviously destructive, they are also an important way of renewing the vitality of a habitat. Some types of plant actually require fires if they are to germinate or reproduce. Empson's poem [third extract] was written after he paid a visit to Kew Gardens. The tree from Turkestan that he mentions seems to have been a pine. The nearby Tree of Heaven that he also mentions was blown down in the great gale of 1987. This latter type of tree does not require a fire as part of its life-cycle. However, because it can regenerate rapidly from any surviving roots, it provides typical early growth after a fire has occurred.

Another specialist area is 'physiological ecology' - how an organism's physiology relates to its environment. Ralph Lewin was a leading authority on marine biology, and spent most of his working life at the Scripps Institution of Oceanography in California. Poetry was his hobby. He particularly liked writing about algae, but as the fourth extract illustrates, he was happy to spread his ecological net wider.

In a sense, ecology has been around for a long time. Some of Gilbert White's notes about the natural history of Selborne in the eighteenth century, for example, could be called ecological. The new ecology that developed in the latter part of the nineteenth century was intended to be both more systematic and more theory-based than the old natural history. Some of the important questions might have been raised by earlier students of natural history, but ecologists now wanted to provide explanations. One example is the balance of nature - how does an ecological community maintain itself? Hudson [fifth extract], though he lived during the early days of ecology, remained an old-style naturalist. (He helped found the Royal Society for the Protection of Birds.) Here he is considering in qualitative terms what is essentially a quantitative question: one that would now be discussed under the heading of *population dynamics*. Stoppard's play [sixth extract] touches on this (and on other scientific themes). The theory referred to here is based on what are technically known as the Lotka–Volterra equations, or, more informatively, as the predator-prey equations. They look at the changes in population that occur when one species preys on another.

It is important to distinguish between ecology and all those activities which now include the word 'green' in their description. Ecologists were, and are, concerned with what is happening, rather than with what ought to happen. Rachel Carson, whose background was in marine biology, is often seen as one of the initiators of our present concern with the quality of the environment. Her book, *Silent spring*, published in 1962, concentrated especially on the adverse environmental effects of synthetic pesticides. In the sixth extract, she imagines what might happen if their use proliferates further. The controversy arising from her book led to restrictions on the use of pesticides and, ultimately, to the creation of the US Environmental Protection Agency. The rapidly mounting interest in the environment has had its influence on poetry. On the one hand, the old 'nature poetry' has developed into 'eco-poetry'; on the other, environmental poetry with a political edge has led to the appearance of 'green poetry' as one way of expressing this concern. Rose Flint's poem [final extract], which appeared in 2008, falls into this latter category, and reflects her wishes for the future of the environment (in this case, in England).

Nature and nature's laws lay hid in night;
God said "Let Newton be" and all was light.
It could not last; the Devil shouting "Ho!
Let Einstein be!" restored the *status quo*.
[J.C. Squire *In Continuation of Pope on Newton*]

In a notable family called Stein
There were Gertrude, and Ep, and then Ein.
Gert's writing was hazy,
Ep's statues were crazy,
And nobody understands Ein.
[Anon.]

freedom
for the daffodils!
——in a tearing wind
that shakes
the tufted orchards——
Einstein, tall as a violet
in the lattice-arbor corner
is tall as
a blossomy peartree
[William Carlos Williams *St. Francis Einstein of the daffodils*]

When he shall feel infuse
His flesh with the rent body of all else
And spin within his opening brain the motes
Of suns and worlds and spaces.
(Einstein enters like a foam)
[Archibald MacLeish *Einstein*]

Space being (don't forget to remember)Curved
(and that reminds me who said o yes Frost
Something there is which isn't fond of walls)
an electromagnetic (now I've lost
the) Einstein expanded Newton's law preserved
conTinuum (but we read that beFore)
[e.e. Cummings *Space being (don't forget to remember) Curved*]

He thought if he could have his space all curved,
Wrapped in around itself and self-befriended,
His science needn't get him so unnerved.
He had been too all out, too much extended.
He slapped his breast to verify his purse
And hugged himself for all his universe.
[Robert Frost *Any size we please*]

Einstein published his theory of general relativity during the First World War. Observations made immediately after the war were found to support the theory. They suggested that Newton's picture of the universe, which had been accepted for over two hundred years, now had to be replaced by Einstein's picture. In consequence, relativity became a buzzword of the 1920s (the period from which all these extracts are taken), and Einstein, himself, became an iconic figure. The first extract is by John Squire, an English writer and critic particularly noted for his epigrams. The one reproduced here is probably his best known. The first two lines were written by Alexander Pope as an intended epitaph for Isaac Newton. Two centuries later, Squire added the second two lines. They reflect the contemporary belief that the Newtonian world picture was a good deal easier to comprehend than its Einsteinian replacement. This same belief appears in the second quotation. (Gertrude Stein was an American writer who spent most of her life in France; Jacob Epstein, noted throughout his life for his controversial sculptures, was also born in America, but became a naturalised British citizen.)

Modernism was a driving force behind much poetry in the 1920s. (The English novelist, Virginia Woolf, even suggested that human nature underwent a fundamental change 'on or about December 1910'.) Its proponents rejected the formal diction and structure adopted by many Victorian poets. The new movement came, it was said, from, 'the modern writer's fervent desire to break with the past, rejecting literary traditions that seemed outmoded and diction that seemed too genteel to suit an era of technological breakthroughs and global violence'. Science came in here as an influence along with technology and war. Einstein's work was seen as overthrowing established science in the same way that modernist poets were overthrowing traditional poetics.

Williams poem - one of the first to mention Einstein - is a good example. It was inspired by Einstein's visit to the United States in 1921. Williams was one of those who wanted to devise a new approach to poetry. He saw Einstein's picture of the universe as supporting his own vision of a world in flux: both the universe outside and the inner world of the individual. (St. Francis is invoked as the patron saint of communication.) Williams had observed the media hype that accompanied Einstein's visit and hoped that modern poetry might likewise be welcomed by American readers.

The next extract is from Archibald MacLeish, another modernist poet in the United States and, later on, the Librarian of Congress. His poem on Einstein attempts to reflect the development of Einstein's life and thoughts. There is seen to be a parallel between changing views of the universe from classical to modern and the corresponding evolution of poetry.

Einstein's general theory of relativity examined how bodies interacted with the space around them. It put forward the concept that the space near a massive body is distorted, so leading to the apparent gravitational attraction of the body. Cummings was both a poet and an artist (his poems often provide visual patterns). Einstein's idea of 'curved space' was guaranteed to interest him. As he says, Einstein's ideas both incorporated, and expanded, Newton's ideas of space. The inclusion of the word 'electromagnetic' may be a reference to the laws of electromagnetism developed by Maxwell in the nineteenth century. Relativity theory needed to incorporate these, as well as gravitation. The poem had a satirical intent. Later in life, Cummings wrote to the editor of a journal: 'please let your readers know that the author of   "Space being(don't forget to remember)Curved" considers it a parody-portrait of one scienceworshipping supersubmoron in the very act of reading (with difficulties) aloud, to another sw ssm'. Cummings slightly misquotes from Frost's poem *Mending wall*. The point he is making is, presumably, that time and gravity pull down walls. Frost's own reflection on Einstein uses the idea of curved space as an encouragement to embrace an apparently indifferent universe. However, he ends up finding it easier to embrace himself.

Salts generally but weakly ..... nor very discoverably by any frication, but if gently warmed at the fire, and wiped with a dry cloth, they will better discover their Electricities.
[Sir Thomas Browne *Pseudodoxia epidemica]*

.... and there a group of girls
In circle waited, whom the electric shock
Dislinked with shrieks and laughter
[Alfred Tennyson *The princess*]

What gave my book the more sudden and general celebrity was the success of one of its proposed experiments .... for drawing lightning from the clouds. This engaged the public attention everywhere. M. de Lor, who had an apparatus for experimental philosophy, and lectured in that branch of science, undertook to repeat what he called the *Philadelphia Experiments;* and, after they were performed before the king and court, all the curious of Paris flocked to see them.
[Benjamin Franklin *Autobiography*]

I was not unacquainted with the more obvious laws of electricity. On this occasion a man of great research in natural philosophy was with us, and, excited by this catastrophe, he entered on the explanation of a theory which he had formed on the subject of electricity and galvanism, which was at once new and astonishing to me.
[Mary Shelley *Frankenstein*]

The lamp-light falls on blackened walls,
And streams through narrow perforations,
The long beam trails o'er pasteboard scales,
With slow-decaying oscillations.
[James Clerk Maxwell *Lectures to women on physical science*]

Here you may put with critical felicity
The following question: 'What is Electricity?'......
Whatever be its nature, this is clear:
The rapid current checked in its career,
Baulked in its race and halted in its course
Transforms to heat and light its latent force
[Hilaire Belloc *Newdigate Poem*]

Lenin used to say that electricity plus socialism equals communism. Our equations are rather different. Electricity minus heavy industry plus birth control equals democracy and plenty. Electricity plus heavy industry minus birth control equals misery, totalitarianism and war.
[Aldous Huxley *Island*]

Now over these small hills, they have built the concrete
That trails black wire; Pylons
[Stanley Spender *The pylons*]

Over the tree'd upland evenly striding,
One after one they lift their serious shapes
[Stanley Snaith *Pylons*]

Electricity   DISCUSSION

Early discussion of electricity centred on static electricity. The Greeks were aware that amber, when rubbed, could attract other objects. Since the Greek word for amber is 'electron', the phenomenon came to be labeled 'electricity' in the seventeenth century. The extract from Sir Thomas Browne appears to be the first use of this word in print. (At the end of the nineteenth century, the particle that carries electrical current was likewise called an 'electron'.) Browne is actually describing experiments to show that amber is not the only substance with this property.

Methods of generating greater quantities of static electricity were developed during the eighteenth century, the most important device being the Leyden jar. In 1750, Abbé Nollet in France arranged a demonstration of its power. He placed a large number of monks in a circle holding hands, and then discharged a Leyden jar through them. To the amusement of the spectators, the monks all leapt into the air simultaneously. The scientific interest was that this meant the electrical discharge must pass very rapidly. The demonstration, however, became an amusement, as Tennyson records in the second extract.

Leyden jars could also discharge through the air, so producing sparks. This led to a comparison with lightning flashes. Two years after Nollet's experiment, Benjamin Franklin carried out an experiment to see whether lightning was simply an electrical discharge. He sent up a kite and used the electricity transmitted down its string to charge up a Leyden jar. As the third extract reflects, this, too, proved to be a popular experiment in France. On the basis of his studies, Franklin proposed that buildings could be protected from lightning strikes by providing a conducting rod from their top to the earth.

In the 1790s, Luigi Galvani in Italy showed that muscles taken from dead frogs could be made to twitch when an electrical discharge was passed through them. He thus demonstrated that nerve impulses have an electrical basis. At the time, the interpretation was less clear. So Mary Shelley [fourth extract] mentions 'electricity' and 'galvanism' - as it came to be called - separately. This ability of electricity to animate dead tissues led to a widespread discussion of the possibility of using it to restore life to the dead. From here, it was but a short step to Frankenstein's monster. James Clerk Maxwell, one of the greatest scientists of the nineteenth century, made an especially important contribution to the study of electricity. He also wrote excellent humorous verse. The fifth extract - a parody of a poem by Tennyson - combines both these interests. Here he is describing the operation of a mirror galvanometer - an instrument for measuring electrical current. Passage of a current causes a mirror to rotate. A spot of light is shone on the mirror; as the mirror rotates, the reflected spot moves along a scale by an amount that depends on the strength of the current. The instrument was used at the time to detect messages transmitted via undersea cables.

The Newdigate prize is offered at Oxford for the best poem submitted by an undergraduate. Hilaire Belloc claimed that his poem, devoted to the joys of electric lighting [sixth extract], was written by a Mr. Lambkin for the 1893 Newdigate prize. Although he extracts considerable amusement from his proposed topic, what he writes about electricity is essentially correct. It is also true that the 1890s were a time when electricity was beginning to make a major impact on the general public (via electric motors, as well as electric lighting). The growing importance of electricity was underlined by the communists soon after the Russian Revolution. Lenin is usually quoted as saying that 'Communism is Soviet power plus the electrification of the whole country'. This is slightly different from Aldous Huxley's version [seventh extract]; but Huxley's main concern, in this his last book, was to paint a different view of the world from that animating his earlier novel *Brave New World*. Yet, as this extract illustrates, in his post-war Utopia electrification still figured as an essential component.

The electrification of the UK began in the 1920s. (The National Grid was set up in 1926 and was in place nationally by 1938.) It became commonplace for the electricity to be conveyed via overhead lines supported by pylons. Many objected to the impact that these pylons had on the landscape; but some left-wing poets revelled in such industrial imagery. Spender was one of these [eighth extract]. Indeed, the group came to be labelled the 'pylon poets', after the poem reproduced here. Snaith's poem [ninth extract] was published about the same time as Spender's in the 1930s. Though Snaith was not one of the 'pylon poets', he similarly saw electricity pylons as a pointer to the future of a changing world.

I am of the opinion that it is not particularly necessary to assert that all planets must be inhabited. However, at the same time it would be absurd to deny this claim with respect to all or even to most of them. Given the richness of nature, where worlds and systems are only sunny dust specks compared to the totality of creation. [Immanuel Kant *Universal natural history and theory of the heavens*]

Whate'er your nature, this is past dispute,
Far other life you live, far other tongue
You talk, far other thought, perhaps, you think, Than man.
[Edward Young *Night thoughts IX*]

The flying figures which came hovering near ..... were the strangest that human eyes had looked upon. In some respects they had a sufficient resemblance for them to be taken for winged men and women, while in another they bore a decided resemblance to birds.
[George Griffith *A honeymoon in space*]

Almost immediately we must have come upon the Selenites. There were six of them, and they were marching in single file over a rocky place, making the most remarkable piping and whining sounds. They all seemed to become aware of us at once, all instantly became motionless, like animals, with their faces turned towards us...... "Insects," murmured Cavor, ""Insects!".
[H.G. Wells *The first men in the Moon*]

[Some worlds] were occupied by races biologically similar to man, others by very different types. The more obviously human races inhabited planets of much the same size and nature as the Earth..... All, whatever the vagaries of their biological history, had finally been moulded by circumstance to the erect form which is evidently most suited to such worlds. Nearly always the two nether limbs were used for locomotion, the two upper limbs for manipulation. Generally there was some sort of head, containing the brain and the organs of remote perception, and perhaps the orifices for eating and breathing.
[Olaf Stapledon *Star maker*]

A further note on this race is that, like those of Deneb III,
Its reproductive method is sexual, which has led
(Relevant at this point) to ability to conceive otherness, mystery,
Illumining life, thought, and especially poem, from the bed.
[Robert Conquest *Excerpt from a report to the Galactic Council*]

It was alien, very alien. It had a peculiar alien tallness, a peculiar alien flattened head, peculiar slitty little alien eyes, extravagantly draped golden robes with a peculiarly alien collar design, and pale grey-green alien skin which had about it that lustrous sheen which most grey-green faces can only acquire with plenty of exercise and very expensive soap.
[Douglas Adams *Life, the universe and everything*]

The official was speaking the human version of Standard, the galaxy's lingua franca. Standard had been chosen as an interspecies, pan-galactic language over eight billion years ago..... [It] was chosen after centuries of research, study and argument by a vast and unwieldy committee composed of representatives from thousands of species, at least two of which became effectively extinct during the course of the deliberations. It was chosen, astonishingly, on its merits, because it was an almost perfect language.
[Iain M. Banks *The Algebraist*]

Extraterrestrial life DISCUSSION

Extraterrestrial life means any life that cannot trace its origins to the Earth: the term is not normally applied to terrestrial life that has been transferred elsewhere. In most writings, the life at the centre of interest is intelligent life, embodiments of such life often being referred to as 'aliens'. In his early days, Kant was much concerned with astronomy [first extract]. He pioneered the idea that the Milky Way is a disc of stars, and that the universe contains other similar stellar discs. It was expected that stars would, like the Sun, be accompanied by planets. In the seventeenth century, it had generally been supposed that all such planets would, like the Earth, acquire life, including intelligent life. As Kant reveals, this belief in the universality of life began to be questioned in the eighteenth century.

Like Kant's *Universal Natural History*, Young's *Night thoughts* appeared around the middle of the eighteenth century. His poem became immensely popular both at home and, in translation, abroad. The ninth section is devoted to natural theology (discerning the existence of God from an examination of the universe). Young firmly concluded that 'an undevout astronomer is mad'. His thoughts on alien life are interesting, for he emphasises its likely differences from human beings [second extract]. There must be, he points out, not just physical differences, but also major differences in the way such beings think and communicate.

Much science fiction has concentrated on the solar system. It has become apparent in recent decades that only the very simplest life forms have much chance of surviving outside the Earth. But in the first half of the last century, it was still just possible that life, even intelligent life might exist on Venus or Mars. The Griffith story [third extract] is an early science-fiction tale dating from the end of the nineteenth century. The author was blithely prepared to find intelligent life at a number of places in the solar system, but his depiction of such life was strongly influenced by ideas from pre-scientific times. Thus the inhabitants of Venus - described in this extract - are attractive and peaceful, whereas the inhabitants of Mars are warlike and barbarian. (The latter are depicted in a similar fashion by both H.G. Wells and Edgar Rice Burroughs.) Even the Moon had not been totally ruled out as a place for life - albeit very simple life - when Wells wrote *The first men in the Moon* in 1901 [fourth extract]. Wells' Selenites (called after Selene, the goddess of the Moon in Greek mythology) exist as a complicated society below the surface of the Moon. C.S. Lewis was inspired to write science fiction by reading Wells, and especially this tale. However, in Lewis's novels, it is his two voyagers from Earth who are evil, while the aliens who are good.

Cavor, Wells' scientist hero, identifies the Selenites as insect-like. Aliens have been imagined as coming in a variety of guises, but they are usually envisaged as the product of the same sort of evolutionary forces that are to be found on Earth. It is widely believed now that life needs Earth-like conditions for it to flourish and develop. As Stapledon explains, if intelligent life exists elsewhere, evolution may lead to a body form similar to that of human beings [fifth extract]. *Star maker* was published in 1937. It was widely read - admired by Wells; disliked by C.S. Lewis. Robert Conquest explores the point made in the second extract by Young: that the basic nature of an organism can affect the way it views the world [sixth extract]. His extraterrestrial observer pinpoints one particular aspect affecting human beings - the influence of their mode of sexual reproduction. Deneb III represents a way of designating the planets moving round a star. In this case, the planet is the third one out from the centre circling the star Deneb. (Deneb, incidentally, is actually not too good a candidate for having planets that support intelligent life.)

'Alien' tends to be a word with negative connotations. So far as extraterrestrial life is concerned, these are reflected in Adams' description of an alien [seventh extract]. The features he mentions are, in fact, a fair description of the Treens: an alien race who figured as the enemies of Dan Dare in the *Eagle*, a popular comic in Adams' youth. He very likely had them in mind, since his first published work appeared in the *Eagle* while he was still at school. One of the continuing questions in science fiction is how different types of alien life might communicate. Adams' unique solution was to invent the 'Babel fish': a small fish which, when inserted in the ear, provided instantaneous translation of any language. Banks (who wrote ordinary novels as Iain Banks and S-F novels as Iain M. Banks) presents a more customary answer in the final extract. The basic supposition is that all the alien species will get together and agree on a common language. It should be added that the message attached to the Pioneer spacecraft, currently heading into interstellar space, was based on graphics and numbers, which were thought to be more universally recognisable than language.

I thought of the coarse white flesh
packed in like feathers,
the big bones and the little bones,
the dramatic reds and blacks
of his shiny entrails,
and the pink swim-bladder
like a big peony.
[Elizabeth Bishop *The fish*]

I know, we Islanders are averse to the belief of these wonders; but there be so many strange creatures to be
now seen, many collected by John Tradescant, and others added by my friend Elias Ashmole, Esq., who now
keeps them carefully and methodically at his house near to Lambeth, near London, as may get some belief
of some of the other wonders I mentioned. I will tell you some of the wonders that you may now see, and
not till then believe, unless you think fit. You may there see the Hog-fish, the Dog-fish, the Dolphin, the
Cony-fish, the Parrot-fish, the Shark, the Poison-fish, Sword-fish,
[Izaak Walton *The compleat angler*]

And when the Salmon seeks a fresher stream to find;
(Which hither from the sea comes, yearly, by his kind,) …..
Here when the labouring fish does at the foot arrive,
And finds that by his strength he does but vainly strive;
His tail takes in his mouth, and, bending like a bow
That's to full compass drawn, aloft himself doth throw,
[Michael Drayton *Poly-olbion*]

Three we kept behind glass,
Jungled in weed: three inches, four,
And four and a half: fed fry to them -
Suddenly there were two. Finally one.
[Ted Hughes *Pike*]

Then there came in a great lazy sunfish, as big as a fat pig cut in half; and he seemed to have been cut in half
too, and squeezed in a clothes-press till he was flat; but to all his big body and big fins he had only a little
rabbit's mouth, no bigger than Tom's.
[Charles Kingsley *The water babies*]

When a fish sickens it's head gets lowest; so that by degrees it stands as it were ont it's head; 'till getting
weaker & losing all poise, the tail turns over; & at last it floats on the water with it's belly uppermost…… It
has been said that the eyes of fishes are immoveable: but these apparently turn them forward or backward in
their sockets as their occasions require. ….. As fishes have no eyelids, it is not easy to discern when they are
sleeping or not, because their eyes are always open.
[Gilbert White *The natural history of Selborne*] Goldfish

"It's a big un. Poke-hooked, too." They hauled together, and landed a goggle-eyed twenty-pound cod. He
had taken the bait right into his stomach. "Why, he's all covered with little crabs," cried Harvey, turning him
over.
[Rudyard Kipling *Captains courageous*]

When a fish is dissected, most of what you see is muscle – which is why they are good to eat [first extract]. Unlike our own muscles, which come in bundles, fish muscles come in easily identifiable layers. They vary in colour from white to red. The white muscles have a relatively restricted blood circulation, and are used when the fish needs to move rapidly. Prolonged, slower swimming relies on the red muscles. The internal organs of a fish differ in colour according to their function. The spleen and the kidney, for example, tend to be dark. The swim bladder deserves its special mention in the first extract. It contains gas and can be controlled to give the fish neutral buoyancy, so that the fish neither sinks to the bottom, nor floats to the surface. Fast-moving fish, such as mackerel, seem to be able to do without it. In some fish, it also appears to be involved in the production or reception of sound.

Izaak Walton is best known for the book from which the second quotation comes. There were two John Tradescants – father and son. Both were, in turn, head gardeners for Charles I. The elder Tradescant was a great collector, and opened the first public museum in England. After his death, the collection was acquired by Elias Ashmole, who subsequently gave it to Oxford University to form part of the basis for the Ashmolean Museum. Some of the fish Walton mentions (such as dogfish) retain their names today. The cony-fish, better known as the burbot, is a freshwater fish. The hog-fish and the parrot fishes are oceanic fishes of the wrasse family. There are various types of poison-fish: some with poisonous spines and others with poisonous flesh. Dolphins are not now counted as fish, but as mammals.

Salmon leaping up waterfalls as they return home to spawn is one of the most remarkable sights that fish can provide. They have been recorded jumping as much as twelve feet into the air. In order to do this, they twist their body like a bent spring: giving the appearance, as Drayton says, of having their tails in their mouths [third extract]. Salmon start their lives in rivers, then swim out to the ocean where they spend their adult lives. When they mature they return to their original river, identifying it apparently by smell. Prior to their run upstream, they undergo internal changes. One, of course, involves development of their sexual organs. The other relates back to our first extract. Jumping requires use of the white muscles, which therefore develop at the expense of the red. Most Atlantic salmon die after spawning, but a few return to the ocean.

Pike have always had a reputation for being aggressive, especially in terms of feeding [fourth extract]. Cannibalism is quite common among them: it occurs when food is scarce, but also if space is limited, for they are territorial animals. Presumably this latter factor was the main problem for Hughes' pike. When pike are large, they will tackle almost anything, but pike a few inches long will only eat smaller fish. We may deduce, therefore, that Hughes' surviving pike was four-and-a-half inches long. The pike is a freshwater fish well-adapted to British rivers. Kingsley's sunfish, on the contrary is a saltwater fish that has drifted far from home [fifth extract]. It prefers warm waters: prolonged periods in water at a temperature of 10°C or lower, common around the British Isles, lead to its death. The sunfish is one of the largest and heaviest of all fishes: its odd mouth is adapted to eating large quantities of jellyfish.

Many fish when they die float belly up. The reason is that their internal organs decay producing gas. The heavier muscles round the spine deteriorate more slowly, so the body inverts with the lighter part on top and the heavier part below. As White says, fish can move their eyes [sixth extract]. Indeed, goldfish, which is what he is discussing in this extract, have been particularly used for studies of eye movements in recent decades. The question of whether fish sleep in the normal sense of the word is much debated. However it is clear that fish do have rest periods when activity is minimal, and this is true of the goldfish that White was studying.

A poke-hooked fish is one that has swallowed the hook down to its belly [final extract]. Atlantic cod have formed the basis for a fishery industry for many centuries. In the latter part of the nineteenth century, when Kipling was writing, the ocean off Newfoundland was the focus for a number of fishing fleets. In recent decades, fishing has become so extensive that cod stocks have plummeted. Though cod attracts a number of parasites, infestation by crabs is unusual.

Golden lads and girls all must,
As chimney sweepers come to dust.
[William Shakespeare *Cymbeline*]

What Beaux and Beauties crowd the gaudy groves,
And woo and win their vegetable Loves……
With honey'd lips enamour'd Woodbines meet,
Clasp with fond arms, and mix their kisses sweet.--
[Erasmus Darwin *The loves of the plants*]

The fragrant Honeysuckle spirals clockwise to the sun
And many other creepers do the same.
But some climb anti-clockwise; the Bindweed does, for one,
Or Convolvulus, to give her proper name.
[Michael Flanders and Donald Swan *Misalliance*]

Of all the propensities of plants, none seem more strange than their different periods of blossoming.
I shall only instance at present in the *crocus sativus* the vernal and the autumnal crocus, which have such an
affinity that the best botanists only make them varieties of the same genus, of which there is only one
species, not being able to discern any difference in the corolla or in the internal structure
[Gilbert White *The natural history and antiquiries of Selborne*]

In buds imprison'd, or in bulbs intomb'd,
    Pervade, PELLUCID FORMS! their cold retreat …..
    From earth's deep wastes electric torrents pour
[Erasmus Darwin The botanic garden, Part I]

        Sir Joseph Banks discreetly
winces under glass.
        And the plants, his plants,
have the last laugh, after all. At Kew, full circle
season and century
[David Malouf *at Kew Gardens*]

From perfect grief there need not be
Wisdom, or even memory:
One thing then learnt remains to me, –
The woodspurge has a cup of three.
[D.G. Rossetti *The woodspurge*]

Not every man has gentians in his house
in Soft September, at slow, Sad Michaelmas.
Bavarian gentians, big and dark, only dark
darkening the daytime torchlike with the smoking blueness of Pluto's   gloom
[D.H. Lawrence Bavarian gentians]

Digitalis as you probably all know occurs in the leaves and seeds of various species of foxglove. But other
plants produce toxic glycosides, too. Strophanthus, oleander, squill and convallaria are some of them.
[Douglas Clark *Vicious circle*]

One of the big problems in discussing flowers is the number of different names that a popular flower can be given, even within a single country. The 'golden lads and girls' of the first extract are dandelions. When they went to seed they were called 'chimneysweepers' in Shakespeare's native Warwickshire (so allowing Shakespeare to make a double pun). The way out of this confusion of multi-names is obviously to accept some kind of generally agreed listing. The system still in use today was introduced by the Swedish scientist, Carl Linnaeus, in the first half of the eighteenth century. ('Linnaeus' was the Latinised version of his name.) He selected the sexual organs of plants as the basis for his classification. His system quickly won wide acceptance, and formed the basis for Erasmus Darwin's poem The loves of plants towards the end of the same century [second extract]. (The poem was published in later editions as part of The botanic garden.) Darwin refers in a footnote to the Linnaean system as involving the 'marriage' of flowers, and mentions the woodbine – a climber that spirals upwards. Such climbers provide the background to a well-known song by Flanders and Swan [third extract]. They say that their song was inspired by an exhibit at the Natural History Museum in London. Most human beings are right-handed; some are left-handed; a few are ambidextrous. Something similar applies to twining plants. Most climb clockwise; some anti-clockwise; a few climb either way. The direction of climbing seems to be genetically controlled, rather than the result of external factors. (Incidentally, by 'woodbine', Darwin probably meant a honeysuckle.)

Gilbert White mentions another characteristic – the blossoming period - that may depend on small genetic differences between apparently similar flowers – in this case, crocuses [fourth extract]. The picture is a little confused here because flowering can also be affected by environmental factors, such as ambient temperature. White's contemporary, Erasmus Darwin, was clear that electricity was another factor affecting growth [fifth extract]. Although experiments looking at the effects of electricity (and magnetism) on germination have continued up to the present day, the results have proved less straightforward to interpret than he supposed.

Sir Joseph Banks, another contemporary of White and Erasmus Darwin, was one of the leading naturalists in Britain [sixth extract]. In the latter part of his life, he was President of the Royal Society (for over forty years), but earlier on he travelled widely – not least with Captain Cook to the South Pacific. Banks became an expert on Australian matters in general, and on its botany in particular. Large numbers of plants were named after him, including the whole Banksia genus. Banks advised George III on the Royal Botanic Gardens in Kew, and dispatched explorers all round the world to look for new plants. The plants were brought back to Kew and cultivated there - if necessary, under glass in greenhouses. In consequence, Kew became one of the leading botanical gardens in the world.

Poets often mention flowers: Wordsworth and daffodils provide an obvious example (though it was apparently Dorothy Wordsworth who drew his attention to them). However, poets do not always get it right. For example, Rossetti, in the seventh extract, ponders on the wood spurge. This happens to have a particularly complex arrangement of flowering parts, so that it is not immediately clear what constitutes the whole flower. However, the only part of it which has three lobes is the stigma, which is clearly not the 'cup'. Lawrence's Bavarian gentians raise a different question [eighth extract]. He specifies 'Michaelmas', which falls on 29 September. But Bavarian gentians only flower from July to August. So it has been speculated that he actually had in mind some other autumn-flowering member of the genus.

Glycosides are a type of organic molecule that contains bound sugar. Plants contain glycosides, some of which can have a significant medical effect on animals and humans. In particular, some glycosides found in flowering plants are highly toxic. A few of these plants, such as the foxglove and the lily-of-the-valley [convallaria], are native to Britain. The others mentioned in the final extract flourish mainly abroad. Their toxicity has been used in actual, or – as here – in fictional murders.

This was a lofty chamber, lined and littered with countless bottles. Broad, low tables were scattered about, which bristled with retorts, test-tubes, and little Bunsen lamps, with their blue flickering flames. There was only one student in the room, who was bending over a distant table absorbed in his work. At the sound of our steps he glanced round and sprang to his feet with a cry of pleasure. "I've found it! I've found it," he shouted to my companion, running towards us with a test-tube in his hand. "I have found a re-agent which is precipitated by haemoglobin, and by nothing else." Had he discovered a gold mine, greater delight could not have shone upon his features.
[Conan Doyle *A study in scarlet*]

The stories have, for the most part, a medico-legal motive, and the methods of solution described in them are similar to those employed in actual practice by medical jurists. The stories illustrate, in fact, the application to the detection of crime of the ordinary methods of scientific research. I may add that the experiments described have in all cases been performed by me, and that the micro-photographs are, of course, from the actual specimens.
[R. Austin Freeman *John Thorndyke's cases*]

The police detective will be as observant as Sherlock Holmes by virtue of his training and years of conditioning. If he cannot match Holmes's brilliant mind, he has learned, through long experience, what to look for and where to look, which is almost as good. And the available laboratory facilities can produce, from clues, information the like of which Conan Doyle would never have dreamed.
[Hillary Waugh *Guide to mysteries and mystery writing*]

'Murder by motor-car,' said Thane. He was more than a little relieved to see Dan Laurence had elected to come along. The Scientific Bureau superintendent ... possessed a needle-sharp mind when it came to the forensic field.... Thane sketched the position in a few sentences. Laurence listened without interruption, then grunted understanding. He turned to his squad, three eager young detective constables, each laden with equipment. 'Right, lads, here we go. Photos first, Willy. Johnny, give him a hand with the flash-gear.... You'll need to take his fingerprints, Frank.'
[Bill Knox *Little drops of blood*]

Rhyme believed in Locard's Principle; in fact, it was the underlying force that drove him to relentlessly push those who worked for him - and to push himself too.
[Jeffery Deaver *More Twisted*]

He took the final step necessary to reach the table and leaned over the remains. There were no identifiable internal organs to be seen ; no heart, no lungs, no liver, no adrenals, no kidneys, only some dry, blackened, anonymous lumps of tissue sticking to the ribs and inner wall of the hide here and there. It was a picture-book stage C4 of the Galloway categorization of mummified remains: "Mummification of tissues with internal organs lost through autolysis or insect activity."
[Aaron Elkins *Skullduggery*]

By five-thirty we were drinking coffee in the convent kitchen, exhausted, fingers, toes, and faces thawing. Elisabeth Nicolet and her casket were locked in the back of the archdiocese van, along with my equipment. Tomorrow, Guy would drive her to the Laboratoire de Médecine Légale in Montreal, where I work as Forensic Anthropologist for the Province of Quebec.
[Kathy Reichs *Death du jour*]

Forensic science - the application of scientific techniques to study legal, especially criminal questions - has made the headlines in recent years. It has, however, a lengthy history, which has been well-reflected in the development of detective stories. Most famously, it forms an important element of the Sherlock Holmes stories at the end of the nineteenth century. Conan Doyle, the author of the stories, attributed his inspiration to Dr Joseph Bell, under whom he worked at Edinburgh Royal Infirmary. Bell believed in the importance of scrutinising patients with care when making a diagnosis. To illustrate his point, he would choose unknown patients and try to deduce their backgrounds and activities from observations alone. This was not, in itself an application of science, but could easily become so, as the Holmes stories illustrate. The initial extract actually comes from the very beginning of the Holmes saga, when Watson meets him for the first time. One of the problems for detection in those days was determining whether or not a stain was actually a blood stain. Here, Holmes is claiming to have found a method for doing this. The story was published in 1887. Chemists had attempted to find tests before this date, but the first fairly reliable test was developed by Dr Kastle in the USA in the early 1900s.

R. Austin Freeman's first detective story appeared before the First World War. His central character, John Thorndyke, has both medical and legal qualifications - not so unusual at time when this was an appropriate background for coroners. Freeman, like Doyle, was a qualified medical practitioner. As the preface quoted here makes clear, he was prepared to develop his own forensic tests. The writings of both Conan Doyle and Freeman helped motivate the developers of forensic science in the twentieth century.

In the third extract, Hillary Waugh outlines the thinking behind 'police procedural' novels. Such stories are concerned with police teamwork, and, not least, with the contribution of forensic investigations. (Waugh, himself, published one of the first detective novels in this genre in 1952.) Increasingly, detailed forensic testing required a team of specialists with access to complex and often expensive equipment.

An example of the team at work - dating from the 1962 - comes next in the extract from Bill Knox. The advent of the police procedural novel signalled difficulties for stories about the lone sleuth who had no access to forensic back-up. These could be circumvented in various ways: often by giving the detective helpful acquaintances within the police force. Thus Deaver's lead character, Lincoln Rhyme, had himself been a major figure in police forensic work before becoming almost totally paralyzed. His assistant remained an active member of the police force. Locard's principle that he mentions can be summed up as 'every contact leaves a trace'. (Locard pioneered forensic science in France, setting up the first police laboratory there in 1910. He was called 'the Sherlock Holmes of France'. Georges Simenon, the detective story writer, attended some of his lectures after the First World War.)

An alternative approach is to make the lead character someone with specialist knowledge who advises the police. Thus Aaron Elkins' hero, Gideon Oliver, is an academic anthropologist who specialises in the study of bones. He travels a good deal, and is consulted by the local police forces when old bones or skeletons are unexpectedly found. In consequence, he has won worldwide renown as the 'Skeleton Detective'. Though mummified remains are less commonly found than ordinary bones, there has been a fair amount of recent research on the topic. Oliver's mention of Galloway refers to some of this work. The ultimate step is to make the hero/heroine a permanent member of an official forensic science team. Kathy Reichs is herself a qualified forensic anthropologist and has worked with police teams in both Quebec and North Carolina. This knowledge is well deployed in stories about her heroine, Temperance Brennan. Both Gideon Oliver and Temperance Brennan have appeared as lead characters in television series.

Armed with hammers, we move along the cliff
Whose blue wall keeps a million million deaths…..Blind
Eyes of belemnites watch from narrow clefts,
Jurassic sun shines on them while they're mined. [Peter Porter *Fossil gathering*]

In the bones of the rock
The fossils are living,
Crinoid and ammonite;
In the red of the rock
(Sandstone and haematite)
The fossils are moving,
Coiling, and crawling,
Aching for the sea. [Norman Nicholson *Fossils*]

Professor Owen ….. takes a fragment of bone, and builds an enormous forgotten monster out of it, wallowing in primeval quagmires, tearing down leaves and branches of plants that flourished thousands of years ago, and perhaps may be coal by this time. [W.M. Thackeray *The Newcomes*]

By one of those familiar conjunctions of things wherewith the inanimate world baits the mind of man when he pauses in moments of suspense, opposite Knight's eyes was an imbedded fossil, standing forth in low relief from the rock. It was a creature with eyes. The eyes, dead and turned to stone, were even now regarding him. It was one of the early crustaceans called Trilobites….. The creature represented but a low type of animal existence, for never in their vernal years had the plains indicated by those numberless slaty layers been traversed by an intelligence worthy of the name. Zoophytes, mollusca, shell-fish, were the highest developments of those ancient dates. The immense lapses of time each formation represented had known nothing of the dignity of man. [Thomas Hardy *A pair of blue eyes*]

The first bed which you will generally find upon the water-worn surface of the chalk is a layer of green-sand and green-coated flints. Among these are met with in many places beds of a great oyster, now unknown in life…… There must have been miles and miles of oyster-bed at the bottom of that Eocene sea.
[Charles Kingsley *Thoughts in a gravel-pit*]

The theory, coarsely enough, and to my Father's great indignation, was defined by a hasty press as being this - that God hid the fossils in the rocks in order to tempt geologists into infidelity…… his reconciliation of Scripture statements and geological deductions was welcomed nowhere, as Darwin continued silent, and the youthful Huxley was scornful, and even Charles Kingsley, from whom my Father had expected the most instant appreciation, wrote that he could not 'give up the painful and slow conclusion of five and twenty years' study of geology, and believe that God has written on the rocks one enormous and superfluous lie'.
[Edmund Gosse *Father and son*]

So careful of the type she seems,
So careless of the single life.
So careful of the type? but no.
From scarped cliff and quarried stone
She cries, "A thousand types are gone;
I care for nothing , all shall go." [Alfred Tennyson *In memoriam*]

Fossils fascinate us for the way they provide knowledge of life on Earth in the distant past. The first two extracts reflect this fascination in the twentieth century. Belemnites looked rather like modern squids, but had an internal skeleton which often remains in fossil form [first extract]. They flourished in the Jurassic period – the age of the reptiles – some 150-200 million years ago. One of the best-known fossil locations in the UK is at Lyme Regis, where belemnites are found in the Blue Lias stratum that colours the cliffs there. (These cliffs also appear in *The French lieutenant's woman*, where they are inspected by the hero, an amateur palaeontologist.) Norman Nicholson's special interest was the Lake District, including its geology [second extract]. Sandstone and haematite [an iron ore] both occur there, though they are not the commonest rocks. The sandstone was laid down when much of Britain was covered by warm shallow seas. Crinoids (multi-armed creatures attached to the bottom) and ammonites (molluscs with spiral shells) – both marine animals - flourished under these conditions.

The detailed study of fossils began in the nineteenth century. Richard Owen was the leading British expert on fossils in that century. He coined the name *dinosaur* and, towards the end of his life, played a major role in the founding of the British Museum (Natural History). He was interested in the classification of animals across the board - vertebrate and invertebrate, fossil and living – and was particularly noted for his ability to extrapolate from incomplete remains [third extract]. He wrote the first definitive paper on belemnites, but was strongly attacked for failing to give credit to its original discoverer. Owen was an important opponent of Darwin's evolutionary ideas in the latter part of the nineteenth century. Thomas Hardy, on the contrary, claimed to be one of Darwin's earliest supporters [fourth extract]. Hardy was interested in geology and acquired a copy of a popular geology text the year before *Origin of species* was published. Trilobites were marine animals with a readily fossilised covering. They were widely distributed and existed over a long period of time. (They were around for over 270 million years, finally disappearing some 250 million years ago.) Trilobites occur as fossils in Cornwall, and it might well be that county which Hardy has in mind here.

The fifth extract comes from a lecture that Charles Kingsley gave to the Mechanics Institute in Odiham, Hampshire, in 1857. Such institutes had originally been founded as a source of education for artisans, not least in science, and many leading educators – Huxley, for example – were happy to give talks there. The lectures often invoked matters of local interest, as Kingsley does here. The Eocene was a period some 30-60 million years ago when part of the South-East of England was covered by a shallow, warm sea. Fossil oysters from this time are a common occurrence. The point to note from this extract is Kingsley's acceptance that life on Earth had a long history. Kingsley, like Hardy, supported Darwin's evolutionary ideas, but did so from within the framework of Anglican theology. One sticking point for some of his contemporaries was the long timescale suggested by fossil history. After all, in the eighteenth century, the Earth was still widely believed to be a few thousand years old. Now an age measured in terms of millions of years was being postulated.

People who took *Genesis* literally were faced with a dilemma. P.H. Gosse provides an interesting example. He was well-known as a naturalist in the nineteenth century – indeed, he was on friendly terms with Darwin – but was also a member of the Plymouth Brethren, a literalist group. After much thought he came up with the proposal – two years before Darwin published *Origin of species* - that the Earth had actually been created in a week with the fossil remains *in situ*. His son's description of life with father is not always reliable, but this particular account is certainly correct. (In passing, Peter Carey's novel *Oscar and Lucinda* is based in part on *Father and son*, including this debate about fossils.)

Fossils raised a further problem for Victorians along with the extended timescale implied. As the Hardy extract notes, many forms of life appeared and disappeared before human beings existed. Tennyson reflects the common Victorian feeling that this battle for life – 'nature, red in tooth and claw', as he puts it elsewhere in this poem – is difficult to reconcile with a loving God. *In memoriam* was written, like Gosse's book, before Darwin's work appeared. The evolutionary implications of fossils had already been spelt out by Robert Chambers in *Vestiges of the Natural History of Creation* published in 1844. (Darwin thought that the book helped to prepare readers for his own work, though he remarked of the author of *Vestiges*, that: 'his geology strikes me as bad, & his zoology far worse'.)

See yonder, lo, the Galaxyë
Which men clepeth the Milky Wey,
For hit is whyt.
[Geoffrey Chaucer *The house of fame*]

Torrent of light and river of the air,
Along whose bed the glimmering stars are seen
Like gold and silver sands in some ravine
[H.W. Longfellow *The Galaxy*]

Until a person has thought out the stars and their interspaces, he has hardly learnt that there are things much more terrible than monsters of shape, namely, monsters of magnitude without known shape. Such monsters are the voids and waste spaces of the sky.
[Thomas Hardy *Two on a tower*]

I noticed, but only to avoid them, great clouds of dust, huge as constellations, eclipsing the star-streams; and tracts of palely glowing gas, shining sometimes by their own light, sometimes by the reflected light of stars. Often these nacreous cloud-continents had secreted within them a number of vague pearls of light, the embryos of future stars.
[Olaf Stapledon *Star maker*]

Across their background glows the Milky Way,
The cradle-place of new born stars untold,
Whose light shall shine adown eternity,
When those now bright have long been dark and cold.
[A. C. Holm *Nox oculis*]

… the galaxy with the 30,000
mph of where
the sun's going:
[A.R. Ammons *Cascadilla Falls*]

But even to us who know how far away
Those constellations burn, the wonder bides
That each vast sun can speed through the abyss
Age after age more swiftly than an eagle,
Each on its different road, alone like ours
With its own satellites; yet, since Homer sang,
Their aspect has not altered!
[Alfred Noyes *Watchers of the sky*]

… because the stellar system to which Terminus belonged was above the Galactic plane, the Galaxy was not seen exactly edge-on. It was a strongly foreshortened double spiral, with curving dark-nebular rifts streaking the glowing edge of the Terminus side. The creamy haze of the nucleus - far off and shrunken by the distance - looked unimportant.
[Isaac Asimov *Foundation's edge*]

The extract from Chaucer is actually the first recorded use of the terms *Galaxy* and *Milky Way* in English. 'Galaxy' derives from the Greek, and Chaucer provides the English name as a translation. The band of light forming the Milky Way bears some resemblance to a stream of milk, and was interpreted that way in ancient mythology. However, as the word 'way' suggests, it could also be interpreted as a heavenly road or river. In England, for example, it was sometimes called 'Watling Street'. A proper understanding of its nature awaited the invention of the telescope. When Galileo used one to examine the Milky Way, he immediately saw that it was actually an assemblage of a large number of faint stars. It is this telescopic picture that Longfellow records in his poem [second extract]. Chaucer gave the word *Galaxy* an initial capital. This practice continues today for a different reason. It is now known that there are many assemblages of stars - all called *galaxies* - in the universe. The use of the initial capital distinguishes our own assemblage from all the rest.

In the latter years of the eighteenth century, William Herschel (best remembered for his discovery of the planet Uranus) carried out a series of detailed telescopic surveys of the Milky Way. He found two things. Firstly, the Sun seemed to sit near the centre of the Galaxy; secondly, there were regions of the Milky Way which appeared to contain very few stars. It was the second of these conclusions that attracted Hardy's attention a century later [third extract]. However, both of Herschel's conclusions came to be questioned in the early decades of the twentieth century. It was realised that there was material lying about between the stars: some was bright gas, but some was dark dust. The dust produced the appearance of dark holes by blocking out the stars behind it. Hardy's 'monsters' did not actually exist. By the 1930s, Stapledon's description of the Galaxy could include both types of interstellar matter. It was also realised that interstellar material and stars were related, in the sense that the latter were born from the former. The bright nebulae (as they were called to distinguish them from the dark nebulae of dust) often marked the spot - as he remarks - where new stars were being born. Holm [fifth extract] makes the same point, but adds that, though some stars are being born, others are dying. The Galaxy is in a continuing state of flux.

Because Herschel overlooked the existence of dust, his estimate of the Sun's position in the Galaxy was also wrong. It was shown in the twentieth century that the Sun is actually a considerable distance - some 27,000 light years - from the galactic centre, which it is orbiting at around 220 km/sec. The figure quoted by Ammons [sixth extract] is considerably less than this. It presumably refers to the Sun's speed relative to its neighbours. All the nearby stars are, like the Sun, orbiting the centre of the Galaxy, but at slightly different speeds. The result is that the Sun drifts relative to nearby stars. The constellations are the two-dimensional patterns traced on the sky by nearby stars. Since both the Sun and the stars are moving relative to each other, the constellation patterns change with time. However, because the stars are far apart, it takes some considerable time for this to become noticeable. As Noyes says [seventh extract], the last three thousand years have been too short a time to produce much change in appearance.

The Milky Way appears as a band in the sky because the Galaxy is a flattened disc within which we sit. Investigations over the past half-century have shown that ours is a spiral galaxy: many of its stars lie in 'arms' that spiral out from the centre. Asimov's *Foundation* series envisage a planet lying at the end of a spiral arm to which the eponymous Foundation is exiled. This planet is, reasonably enough, named *Terminus*. The final extract describes the view of the Galaxy from Terminus. Whereas the Sun is situated near the central plane of the galactic disc, the star that Terminus orbits lies some distance above that plane. Consequently, an observer can look down on the disc and see the clouds of bright and dark nebulae that trace the path of the spiral arms. The central nucleus of the Galaxy is actually a prominent bulge, but at the distance of Terminus - considerably further from the centre than the Sun - it is diminished. When Asimov wrote the *Foundation* series, the Galaxy was thought to be an ordinary spiral galaxy with its arms starting out at the central bulge. More recent observations suggest that we may be living in a barred spiral galaxy. In such a galaxy, a bar of stars sticks out from the central nucleus, and the spiral arms emanate from the ends of the bar.

The Pythagoreans were bound together in a brotherhood, the members of which had rules that are not now understood, but which linked them so as to form a sort of club, with common religious observances and pursuits of science, especially mathematics and music.
[Charlotte M. Yonge *The two friends of Syracuse*]

As great Pythagoras of yore,
Standing beside the blacksmith's door,
And hearing the hammers as they smote
The anvils with a different note,
Stole from the varying tones that hung
Vibrant on every iron tongue,
The secret of the sounding wire,
And formed the seven-chorded lyre.
[Henry Wadsworth Longfellow *Verses to a child*]

That Greek one, then, is my hero who watched the bath water
Rise above his navel, and rushed out naked, 'I found it,
I found it' into the street
[Danny Abse *Letter to Alex Comfort*]

To Archimedes once a scholar came,
"Teach me," he said, "the art that won thy fame;--
The godlike art which gives such boons to toil,
And showers such fruit upon thy native soil;--
The godlike art that girt the town when all
Rome's vengeance burst in thunder on the wall!"
[Friedrich von Schiller Translated from *Archimedes*]

...I've not the least doubt
That he'd placed up his sleeve
Mr Todhunter's excellent Euclid,
The same with intent to deceive
[A.C. Hilton *The heathen passee*]

And therefore in geometry (which is the only science that it hath pleased God hitherto to bestow on mankind), men begin at settling the significations of their words; which settling of significations, they call definitions, and place
them in the beginning of their reckoning.
[Thomas Hobbes *Leviathan*]

He studied and nearly mastered the Six-books of Euclid (geometry) since he was a member of Congress. He began a course of rigid mental discipline with the intent to improve his faculties, especially his powers of logic and language. Hence his fondness for Euclid, which he carried with him on the circuit till he could demonstrate with ease all the propositions in the six books; often studying far into the night, with a candle near his pillow, while his fellow-lawyers, half a dozen in a room, filled the air with interminable snoring.
[Abraham Lincoln *Short autobiography*]

The Renaissance, as the name implies, marked a revival of the learning that originated in the classical world. Importantly for the development of science, this revival included mathematics. One of the earliest of the Greek mathematicians to receive attention was Pythagoras (who lived in the 6th century BC). Pythagoras is best remembered for his theorem relating to the squares on the sides of a right-angled triangle. As the first extract reminds us, he established a quasi-religious brotherhood in South Italy. In consequence, it is difficult to tell which of the mathematical ideas were his, and which came from his disciples. Either way, the Pythagoreans helped initiate the study of geometry. They were also concerned with numbers, and Pythagoras is said to have discovered that harmony comes from strings whose lengths are in simple ratios. The folk tale regarding this discovery is retold in the second extract. These ideas led to further developments. For example, the number of notes in a harmonic scale was similar to the number of celestial bodies that were believed to orbit the Earth. Hence grew the idea of 'the music of the spheres', of which Shakespeare was so fond. (The 'spheres', in this case, means the orbits.) More importantly, it became customary to associate geometry and astronomy, on the one hand, and numbers - that is arithmetic - and music, on the other. These four formed the quadrivium, the basis of more advanced teaching throughout the Renaissance.

Archimedes lived three centuries after Pythagoras, and is widely regarded as the greatest mathematician of antiquity. (Indeed, he is often ranked alongside Newton and Einstein in terms of achievement.) The story most people know about Archimedes relates to his discovery of a method for finding the volume of an object with an irregular shape. The story goes that he realised, when taking a bath, that the volume of his body could be measured by the amount of water that he displaced. He was so excited that he ran into the street naked shouting, 'Eureka' (meaning, as the third extract says, 'I have found it' - though Abse was, perhaps, more concerned with the nudity than with the discovery). Archimedes was an engineer, as well as a mathematician, and worked in this capacity in his home city of Syracuse. His inventions covered both civil and military activities. The extract from Schiller refers to stories of how he helped in the defence of Syracuse against the besieging Romans. He was killed - apparently against the orders of the officer in charge - when the Roman forces finally captured the city.

Euclid lived some years before Archimedes. Little is known of the man, but his *Elements* has been one of the most influential textbooks in the history of mathematics. It is really a compendium of geometry, bringing together all that was known in his time. It was used in teaching into the latter part of the twentieth century. Various mathematicians produced editions of the work; the best-known in Britain was that produced by Isaac Todhunter in the mid-nineteenth century. (Todhunter was a Cambridge don: his pupils included Leslie Stephen, the father of Virginia Woolf.) Many a student found Euclid hard going. One contributor to the magazine *Punch* wrote a poem on the death of Euclid declaiming: 'This is he/ who with his learning made our youth a waste'. Hilton's heathen passee at Cambridge certainly felt the same. (The poem, incidentally, is a parody of Bret Harte's *The Heathen Chinee*, which was itself a parody of Swinburne).

Euclid's *Elements* was a mainstay of teaching because it was seen as an exemplar of how to think logically. Starting from a few basic postulates, it leads on to a series of propositions which necessarily follow. This approach has had a continuing influence on philosophy. An obvious example is Thomas Hobbes. According to John Aubrey, Hobbes was already forty years old before he became interested in geometry; being in a library one day, he casually took up a copy of the *Elements* and became fascinated by its contents. The consequences for his thinking are reflected in the penultimate extract. Geometry, as the name implies, was linked to making measurements of the Earth's surface. It is thus of basic importance for the work of surveyors, a profession which played a vital role in the expansion of the USA. (Thus George Washington, while still a young man, was appointed the Surveyor-General for Virginia.) As the final extract shows, however, Euclid was also valued, as in Britain, for its virtue in training the mind. The *Elements* actually appeared in thirteen books. The first six covered elementary geometry, and these were the ones normally used for teaching purposes. All the extracts quoted here are referring to these six books.

Heat is a motion; expansive, restrained, and acting in its strife upon the smaller particles of bodies. But the expansion is thus modified; while it expands all ways, it has at the same time an inclination upward. And the struggle in the particles is modified also; it is not sluggish, but hurried and with violence.
[Sir Francis Bacon *The new organon*]

Close as I ever came to seeing things
The way the physicists say things really are
Was out on Sudbury Marsh one summer eve
When a silhouetted tree against the sun
Seemed at my sudden glance to be afire:
A black and boiling smoke made all its shape.
[Howard Nemerov *Seeing things*] Evaporation,
    that random breach of surface tension
by molecules "which happen to acquire exceptionally high
    velocities."
[John Updike *To evaporation*]

What happens in these Lattices when *Heat*
Transports Vibrations through a solid mass? ....
Debye in 1912 proposed Elas-Tic Waves called phonons.
[John Updike *The dance of the solids*]

listening deeply, we might fancy
infinitesimal clicks
as each tailored wafer builds
its strict array [Alice Fulton *Disorder is a measure of warmth*]

1. You can't win, you can only break even.
2. You can only break even at absolute zero.
3. You can never reach absolute zero.
{Anon. *Summary of the laws of thermodynamics*]

Till in that twilight of the gods
When earth and sun are frozen clods,
When, all its matter degraded,
Matter in aether shall have faded,
[James Clerk Maxwell *A Paradoxical Ode*]

 'Such is life,' although rarely is it described in this manner: an inserting itself, a drawing off to its advantage, a parasitizing of the downward course of energy, from its noble solar form to the degraded one of low-temperature heat. In this downward course, which leads to equilibrium and thus death, life draws a bend and nests in it. [Primo Levi *The Periodic Table*]

A good many times I have been present at gatherings of people who, by the standards of the traditional culture, are thought highly educated and who have with considerable gusto been expressing their incredulity at the illiteracy of scientists. Once or twice I have been provoked and have asked the company how many of them could describe the Second Law of Thermodynamics. The response was cold: it was also negative. Yet I was asking something which is the scientific equivalent of: Have you read a work of Shakespeare's?
[C.P. Snow *The Two Cultures and the Scientific Revolution]*

The new organon, which appeared in 1620, was written as a part of Bacon's argument for a new way of studying nature [first extract]. He argued for the importance both of induction and experiment. One of his interests was heat (or cold as it was often called in the seventeenth century). As this extract suggests he associated heat in a general way with the motion of particles. However, until the nineteenth century there was a parallel belief that heat was a fluid which, by flowing from place to place, determined the temperature distribution. The idea that gases were assemblages of small particles in constant motion and colliding with each other was developed in the eighteenth century, but a full quantitative theory of this kinetic theory, as it was called, only occurred in the latter part of the nineteenth century. Nemerov sees a cloud of gnats as analogous to such a gas as visualised by a physicist [second extract].

Materials can undergo changes of state: from solid to liquid or liquid to gas. Such changes require adding or removing heat. When heat is added to a liquid the molecules it contains move around more rapidly. But there is a distribution of speeds – some molecules move more rapidly than others. Faster moving ones, if they reach the surface of the liquid, can leave it before actual boiling occurs. The process is called evaporation [third extract]. (Surface tension is simply a result of the attraction of the molecules of the liquid for each other, which tries to prevent them leaving.)

In solids the atoms are fixed in position, but are still capable of vibrating. The amount of vibration depends on the temperature. Because the atoms interact with each other, the vibrations combine collectively throughout the solid as waves. The amount of energy associated with each vibration is called a phonon. This picture of how heat is transmitted through a solid was first developed in detail – as Updike says [fourth extract] – before the First World War by Peter Debye, a Dutch physicist. Updike's poem looks at various kinds of solids. The sort of solid Debye had in mind in formulating his theory was the simplest of all - a regularly ordered and homogeneous crystal. Low frequency vibrations in solids produce the equivalent of sound waves (indeed, *phonon* comes from the Greek word for 'sound'). Alice Fulton is writing about frost, in other words ice crystals, which are examples of this simple solid. She imagines that, as the ice crystal forms, each successive layer produces a slight sound as it is added to the crystal [fifth extract].

The temperature of a system relates to its energy. During the nineteenth century, the study of heat and energy came to be labelled *thermodynamics*. Work during that century led to the formulation of two laws of thermodynamics. The first says that energy can be transformed, but not created. The second says, in effect, that in any transformation some energy is always dissipated as heat. The idea of an absolute zero temperature was also introduced in the nineteenth century. Since temperature is a measure of how fast molecules move, the lowest temperature possible should occur when all the molecules are stationary. The third law of thermodynamics – that absolute zero is inaccessible - was formulated in the early twentieth century. Taken together, the three laws reveal that all types of energy must ultimately degrade into heat [sixth extract].

The second law had a major impact on nineteenth-century thought. It implied that the universe would ultimately become a changeless sea of material all at a uniform temperature. This is the picture that Maxwell presents: it came to be called the 'heat death' of the universe [seventh extract]. Nineteenth-century scientists believed the universe was pervaded by an invisible substance – the aether – so Maxwell further speculates that matter will eventually merge in with the aether. (The idea of such an aether was dropped in the twentieth century.) In the path towards this heat death, it is always possible to divert some of the remaining energy to create something new. Levi notes that life on Earth survives in this way by using some of the energy from the Sun before it dissipates [eighth extract].

C.P. Snow, who trained as a physicist, famously used the second law of thermodynamics to illustrate his proposed division of intellectuals into two cultures – one involving science and the other the humanities [final extract]. He pointed out that the division was asymmetric: scientists usually knew something about the humanities, but those immersed in the humanities were often distinctly limited in their knowledge of science.

I shall confine myself, in discussing this question, to those fragmentary Human skulls from the caves of Engis in the valley of the Meuse, in Belgium, and of the Neanderthal near Düsseldorf, the geological relations of which have been examined with so much care by Sir Charles Lyell; upon whose high authority I shall take it for granted, that the Engis skull belonged to a contemporary of the Mammoth (*Elephas primigenius*) and of the woolly Rhinoceros (*Rhinocerus tichorhinus*), with the bones of which it was found associated; and that the Neanderthal skull is of great, though uncertain, antiquity.
[T.H. Huxley *Man's place in nature*]

'Speak, O man, less recent! Fragmentary fossil!
Primal pioneer of Pliocene formation,
Hid in lowest drifts below the earliest stratum
Of volcanic tufa! [Francis Bret Harte *To the Pliocene skull*]

At length as an ape he was fain
The nuts of the forest to rive,
Till he took to the low-lying plain,
And proceeded his fellows to knive.
[Grant Allen *The lower slopes*]

From age to dumb unnumbered age,
By dim gradations long and slow,
He reaches on from stage to stage,
Through fear and famine, weal and woe
And, compassed round with danger, still
Prolongs his life by craft and skill.
[Matilda Blind *The ascent of man*]

….. I'll try a heat recipe.
Stone-bashing strikes my thrifty simile
for fire-making for when two flints collide
sparks fly
[Peter Howe *The ascent of man*]

For a century or so, the scientific problem of the origin of man has been under discussion, and a swelling team of research workers has been digging feverishly into the past to discover the initial point of hominisation, and yet I cannot find a more expressive formula than this to sum up all our prehistoric knowledge. The more we find of fossil human remains and the better we understand their anatomic features and their succession in geological time, the more evident it becomes, by an unceasing convergence of all signs and proofs, that the human 'species',
however unique the ontological position that reflection gave it, did not, at the moment of its advent, make any sweeping change in nature.
[Pierre Teilhard de Chardin *The phenomenon of man*]

I call them apes, but they carried sticks and stones in their hands and jabbered talk to each other, and ended up by tyin' our hands with creepers, so they are ahead of any beast that I have seen in my wanderin's. Ape-men - that's what they are - Missin' Links, and I wish they had stayed missin'.
[A. Conan Doyle *The lost world*]

Human evolution   DISCUSSION

The growth of interest in geology and archaeology during the nineteenth century led to the discovery of an increasing number of human remains from ancient times. In 1823, the Oxford geologist, William Buckland, found the so-called Red Lady of Paviland in a Welsh cave. He thought it dated from Roman times. (Subsequent studies have shown it to be a male from perhaps 30,000 years ago.) This skeleton was clearly similar to a modern human being; but some remains were more unusual. Skulls discovered in Belgium in 1829, Gibralter in 1848, and Germany in 1856 showed unusual features. It is these skulls that Huxley is discussing [first extract]. Charles Darwin's *On the origin of species* appeared in 1859. Huxley's book, published in 1863, applied Darwin's theory of evolution to human beings. During the 1860s, these skulls were recognised as belonging to a different human species - Neanderthal Man. (The German specimen was found in the Neander Valley.) Huxley's references to the mammoth and the woolly rhinoceros - both extinct - were designed to show how old the Neanderthals were. The Pliocene period ended even before this, and really old hominid fossils had not been found when Bret Harte was writing [second extract]. However, various human bones had been dug up as a result of the Calfornia gold rush in the mid-nineteenth century. Bret Harte, who moved to California around that time, picked up speculations that they were very old from press reports. He clearly did not take these reports too seriously: the bones in the poem are revealed as those of a recent miner.

Huxley examined in his book how primates differed from other mammals. By the early twentieth century, it was generally agreed that one important difference lay in the way primates were adapted for living in trees. It was suggested that this led to such vital human characteristics as grasping hands and keen vision. It was consequently proposed that a significant step in the evolution of humans was when their ancestors came down from trees, bringing these physical characteristics with them. Grant Allen was a Canadian who settled in England in the latter part of the nineteenth century. He became a convinced Darwinian, and in the third extract (published in 1894) succinctly summarises what was known as the 'arboreal hypothesis' relating to human origins.

Less importance is attached to tree-dwelling nowadays, but another aspect, then also under discussion - man as toolmaker - continues to be important. From early on, hominid remains were found in association with stone tools. The sophistication of these tools gradually developed down the millennia. This was presumably what Mathilde Blind had in mind when she referred to 'craft and skill' [fourth extract]. Blind, a German who settled in England, wrote this poem as a tribute to Darwin's theory of evolution. Another significant development was the ability to control fire. Peter Howe [fifth extract] associates this with the use of flints to strike sparks. His poem is divided into four stanzas, each devoted to a different hominid. The section quote here is headed *Pithecanthropus*. (The name translates as 'ape-man'.) The first fossils of *Pithecanthropus* were found near the end of the nineteenth century, but the association of the remains with fire is a more recent discovery. It is interesting that two poets a century apart should entitle their poems 'The ascent of man', whereas Darwin's book on human evolution was called 'The descent of man'.

Pierre Teilhard de Chardin was both a palaeontologist and a Jesuit priest. He is best remembered now as a philosopher with a special interest in evolution [sixth extract]. In the 1930s, he was involved in the discovery in China of an early hominid labelled Peking Man. Some years prior to this, he was involved in a much more controversial discovery - Piltdown Man. In the period just before the First World War, bone fragments were discovered at Piltdown in Sussex which seemed to belong to a hominid with a modern skull, but with an ape-like jawbone. Doubts were soon cast on this interpretation, but it was not until after the Second World War that the whole thing was shown to be a deliberate hoax - a mediaeval human skull, the jawbone of an orang-utan, and chimpanzee teeth. (The hoax is usually attributed to Teilhard de Chardin's friend, Charles Dawson, rather than to he, himself.) One reason why the hoax was so successful was because, in the light of Darwin's ideas on evolution, a major search was then going on for 'missing links' (fossils showing the evolutionary transition from one group of creatures to another.) As Conan Doyle indicates [seventh extract], the term came to be attached especially to human evolution. *The lost world* appeared in 1912, the year in which the Piltdown skull was unearthed, but, if Conan Doyle had a particular missing link in mind, it would have been *Pithecanthropos*. As noted above, remains of this hominid had been discovered just over a decade before Conan Doyle's story appeared.

These tiny loiterers on the barley's beard,
And happy units of a numerous herd
Of playfellows, the laughing Summer brings,
Mocking the sunshine on their glittering wings,
How merrily they creep, and run, and fly!
[John Clare *Insects*]

The Creator would appear as endowed with a passion for stars, on the one hand, and for beetles on the other, for the simple reason that there are nearly 300,000 species of beetle known, and perhaps more, as compared with somewhat less than 9,000 species of birds and a little over 10,000 species of mammals. Beetles are actually more numerous than the species of any other insect order.
[J.B.S. Haldane *What is life?]*

The most insignificant insects and reptiles are of much more consequence, and have much more influence in the economy of nature, than the incurious are aware of and are mighty in their effect, from their minuteness, which renders them less an object of attention, and from their numbers and fecundity.
[Gilbert White *The natural history of Selborne*]

….. met with the Comptroller at the office a little both forenoon and afternoon, and at night step a little with him to the Coffee House where we light upon very good company and had very good discourse concerning insects and their having a generative faculty as well as other creatures.
[Samuel Pepys *Diary 1660*]

It is a significant fact, stated by entomologists - I find it in Kirby and Spence - that "some insects in their perfect state, though furnished with organs of feeding, make no use of them"; and they lay it down as "a general rule, that almost all insects in this state eat much less than in that of larvae. The voracious caterpillar when transformed into a butterfly….. and the gluttonous maggot when become a fly" content themselves with a drop or two of honey or some other sweet liquid.
[Henry Thoreau *Walden*]

Death is like the insect
Menacing the tree,
Competent to kill it,
But decoyed may be.

Bait it with the balsam,
Seek it with the saw
[Emily Dickinson *Death is like an insect*]

V. B. Wigglesworth wakes at noon,
Washes, shaves and very soon
Is at the lab; he reads his mail,
Swings a tadpole by the tail,
Undoes his coat, removes his hat,
Dips a spider in a vat
Of alkaline
[John Updike *V. B. Nimble, V. B. Quick*]

Writing about nature was a popular activity in the nineteenth century. Often such writings mentioned specific insects, especially bees, but insects as a group sometimes appeared - as, for example in Wordsworth Amongst the poets, however, the most knowledgeable was John Clare, at least as concerns the insects of his native Northamptonshire [first extract]. Clare could be relaxed about the variety of insects on the barley, since barley was one of the less disease-ridden food plants in the nineteenth century.

J.B.S. Haldane, as someone devoted to biological statistics, saw the variety of insects in a different light. In terms of number of species identified, insects appear to dominate the Earth. Within the various insect groups, beetles - as he was one of the first to emphasize in popular writing - dominate the remainder [second extract]. There is a famous, but probably apocryphal story that he was asked what could be inferred about God from a study of the natural world. He is alleged to have replied that God must have the characteristics of a beetle in view of his inordinate fondness for them. Though Haldane could quote statistics, the important role that insects played in nature had been identified long before. Insects had been studied from antiquity, but it was mainly the more obvious groups - such as bees and butterflies - that attracted attention. By the latter part of the eighteenth century, the vast number of different insect species was becoming realised [third extract]. (In those days, the word 'reptile' could mean any creeping or crawling creature.) That period was significant because it laid down the principles that have ever since guided insect classification. But it was also the time when the economic impact of insect activities began to attract attention.

Interest in the anatomy of insects had grown in the seventeenth century: not least because of the invention of the microscope. The most famous drawing in Hooke's *Micrographia*, published in 1665, is of a flea, but he included representations of other insects, too. Only a few years before, we find Pepys relaxing in a coffee-house by discussing the reproduction of insects [fourth extract]. This was a live topic at the time. The first accurate description of the reproductive organs of insects was published by the microscopist, Jan Swammerdam, in 1669. By the nineteenth century, the way in which insects lived was also a focus of attention. Thoreau writing on their feeding habits [fifth extract] provides an example. The *Introduction to entomology* by the Reverend William Kirby and William Spence, that Thoreau mentions, was published in England early in the century and went through numerous subsequent editions. It was a standard reference source throughout the nineteenth century. The word 'entomology', itself, had come into use in the latter part of the eighteenth century.

The growth of agriculture in the nineteenth century - especially where large areas were devoted to the same crop - emphasised the importance of insects as pests. The late eighteenth and early nineteenth century saw the appearance of the first publications on pest control, and these rapidly increased in number from then on. In the north-east of America in the latter half of the nineteenth century, large areas of spruce trees were killed off by insects that bored into them. This was something of which Emily Dickinson would have been aware. The standard advice to counter these depredations was to cut down the infested timber and take it away. 'Seek it with a saw' as Emily Dickinson says [sixth extract]. An alternative was to coat the threatened trees with some form of insect repellent. In the nineteenth century, artificial pesticides were in their infancy: most use was made of naturally occurring substances. One of those used against insect pests was a particular form of balsam, also known as 'tree oil'. ('Decoyed' here actually means 'averted'.)

Studies of insects initially concentrated on their anatomy and classification. By the end of the nineteenth century, attention was turning to their physiology - how they worked. The leading exponent of insect physiology in the twentieth century was Vincent Wigglesworth (he was later knighted, and became Sir Vincent). In the 1950s, he was appointed to the Quick Chair of Biology at Cambridge University. 'Quick' was the name of the man who endowed the chair, but the *New Yorker* thought it appropriate for Wigglesworth in view of his unending stream of publications. This point, along with the name 'Wigglesworth' for an expert on insects, greatly appealed to John Updike and led to the poem quoted in the final extract.

Grind away, moisten and mash up thy paste,
Pound at thy powder, -- I am not in haste.....
That in the mortar—you call it a gum?
Ah, the brave tree whence such gold oozings come!
[Robert Browning *The laboratory*]

Jakob didn't return to the laboratory table to complete the measurement, which could wait. Instead, he set off to visit the Geheimrat, whom he used to find down among the fragile and costly pumps and glassware in the low-temperature laboratory, but more often than not he now found in his office.
[Russell McCormmach *Night thoughts of a classical physicist*]

The biological laboratory had an atmosphere that was all its own. It was at the top of the building, and looked clear over a clustering mass of inferior buildings toward Regent's Park. It was long and narrow, a well-lit, well-ventilated, quiet gallery of small tables and sinks, pervaded by a thin smell of methylated spirit and of a mitigated and sterilized organic decay. Along the inner side was a wonderfully arranged series of displayed specimens that Russell himself had prepared.
[H.G. Wells *Ann Veronica*]

I passed all the other courses that I took at my University, but I could never pass botany. This was because all botany students had to spend several hours a week in a laboratory looking through a microscope at plant cells, and I could never see through a microscope. I never once saw a cell through a microscope. This used to enrage my instructor.
[James Thurber *My life and hard times*]

Arthur had been blowing glass in the Haughton Laboratory for forty-five years. It was his favourite little joke when demonstrating to students. 'Professor Sewell and me came to the Haughton about the same time, him as an undergraduate and me as a lab. boy. Professor Sewell went on and studied physiology and zoology and the biochemistry and I went on and studied glass. We both learnt a lot since then, gentlemen.'
[Nigel Balchin *A sort of traitors*]

Working in the laboratory
You're exact, if nothing else.
I like my white-coated sterility.
[Eva Royston *Working in the laboratory*]

The lab was furnished with worn wooden benches, set close together to make the most of available space. Above the benches were old-fashioned gas and electrical outlets, the latter festooned untidily, and probably unsafely, with adapters and many plugs. On the walls were roughly-made shelves, all filled to capacity with books, papers and apparently discarded equipment
[Arthur Hailey *Strong medicine*]

The laboratory in which they talked (Gottlieb pacing the floor, his long arms fantastically knotted behind his thin back; Martin leaping on and off tall stools) was not in the least remarkable - a sink, a bench with racks of numbered test-tubes, a microscope, a few note-books and hydrogen-ion charts, a grotesque series of bottles connected by glass and rubber tubes on an ordinary kitchen table at the end of the room.
[Sinclair Lewis *Arrowsmith*]

The *OED* defines the old word 'elaboratory' as meaning 'a place where chemical operations are performed, or where medicines are compounded'. Laboratories, as we know them today, are a nineteenth-century invention; but they evolved out of this need for a place to carry out chemical operations. Liebig developed the first modern laboratory - designed to train chemists - during the 1830s at Giessen in Germany. Browning's poem [first extract] was published in the following decade. Though it uses the word 'laboratory', the setting is actually seventeenth-century France. The words quoted here are spoken to an apothecary by a woman for whom he is preparing a poison. By this time, some of the apparatus that later typified laboratories was already in use. This extract mentions the mortar (which is always accompanied by a pestle). The mortar, a tough bowl, and the pestle, an equally tough club, are used to grind up substances as finely as possible in the preparation of chemicals or medicines. They have been commonplace for so long that they have sometimes been called 'apothecary grinders'.

By the latter part of the nineteenth century, laboratories were an integral part of all the experimental sciences. The layout of the laboratory was roughly similar for all of them, but the apparatus varied. The second extract is set in a German physics laboratory early in the twentieth century. In such a laboratory, the provision of electricity and of electrical apparatus was more important than in a chemical laboratory. Here, the main concern was with the apparatus required for low-temperature physics. This was then a new field of research. In the decade just before the setting of this story, Onnes at Leiden had succeeded in liquefying the newly discovered gas, helium, reaching temperatures only five degrees above absolute zero. (In passing, German university staff were - and are - civil servants, so senior staff might also hold state appointments - a *Geheimrat* is a councillor.)

The contents of biology laboratories differed yet again - most obviously [in the third extract] by the biological specimens prepared by Russell (modelled by Wells on T.H. Huxley). The smell of methylated spirits was pervasive because it was then widely used for preserving specimens and tissues. The most obvious difference, however, was the prevalence of microscopes, which were used to examine the preserved tissues. As Thurber [fourth extract] makes clear, the ability to use a microscope was regarded as essential for a biologist, which could sometimes present a problem for people with defective eyesight.

In addition to academic staff, there has been an increasing need for laboratory assistants to help set up experiments, and for technicians to help build apparatus (though, in earlier days, one person might combine both activities). In chemistry and physics laboratories, in particular, there was often a need for glass apparatus that could not be bought off the shelf. This was made in-house by a technician specially skilled in glass-blowing techniques - like the one in the fifth extract. As this extract indicates, a skilled technician might also be involved in the laboratory, teaching basic techniques to students.

The white coat has come to be seen as synonymous with the scientist and the laboratory [sixth extract]. Its use dates back to the nineteenth century. It has two main purposes: firstly, to protect the clothes against accidents - for example, spillage of chemicals - which is why it is an enveloping coat; secondly, to ensure cleanliness - which is why it is white, to allow stains to be seen. Hospital doctors have adopted similar garb for the same reasons (and perhaps also as a distinguishing uniform).

The previous extracts (apart from the first) have been mainly about teaching laboratories, where students learn about the practical side of their science. University researchers have increasingly carried out their own research in laboratories where none below the level of doctoral students may venture. In times past, when laboratories might well be sited in buildings not necessarily designed for modern research, the result could look chaotic [seventh extract]. Laboratories in research institutes, or attached to commercial firms, might contain similar apparatus, but were usually designed for the purpose and kept in a more orderly state [eighth extract]. The laboratories in *Arrowsmith* were modelled on those at the Rockefeller Institute for Medical Research in New York in the 1920s. Needless to say, they were both adequately designed and well-funded.

Music is the arithmetic of sounds as optics is the geometry of light.
[Claude Debussy   Quoted by Dean Keith Simonton *Greatness*   *: Who Makes History and Why*]

.... 'twere not absurd
To doubt, if beams, set out at Nature's birth,
Are yet arrived at this so foreign world;
Though nothing half so rapid as their flight.
[Edward Young *Night thoughts: Night ninth*]

Nor could the darting beam of speed immense
Escape his swift pursuit and measuring eye.
Ev'n Light itself, which every thing displays,
Shone undiscover'd, till his brighter mind
Untwisted all the shining robe of day;
And, from the whitening undistinguish'd blaze,
Collecting every ray into his kind,
[James Thomson *A poem sacred to the memory of Sir Isaac Newton*]

The sun, the wind, bring rain
And I know what the rainbow writes across the east or the west in a half-circle
[Carl Sandburg *Prairie*]

He then, in a strain of humour beyond description, abused me for putting Newton's head into my picture; " a fellow," said he, "who believed nothing unless it was as clear as the three sides of a triangle." And then he and Keats agreed he had destroyed all the poetry of the rainbow by reducing it to the prismatic colours.
[Benjamin Haydon   *An immortal evening*]
As to what I have done as a poet... I take no pride in it... but that in my century I am the only person who knows the truth in the difficult science of colours – of that, I say, I am not a little proud, and here I have a consciousness of a superiority to many.
 [Johann Goethe Quoted by Johann Eckermann *Conversations of Goethe*]

As for the lights they hang about the town,
Some praise them highly, others run them down.
This system (technically called the Arc),
Makes some passages too light, others too dark.
[Hilaire Belloc *Newdigate poem*]

The atoms clash, the spectra flash,
Projected on the screen,
The double D, magnesian b,
And Thallium's living green.
[James Clerk Maxwell   *To the Chief Musician upon Nabla*]

Sometimes (the legend tells)
Those dervish dancers leap
Their minute void across …..
But in that waste between
That 'here', that 'there'
No path can be
Nor compromise [H. Witheford *Bohr on the atom*]

In the classical curriculum of the quadrivium, music – as Debussy says – was linked to arithmetic; but geometry was linked to astronomy, rather than to optics [first extract]. However, the study of how light reflects from mirrors, or passes through lenses, is, indeed, often called 'geometrical optics'. In the 1670s, the Danish astronomer, Ole Rømer, showed from observations of Jupiter's moons that light travels at a finite speed. As Young, writing in the following century, notes, this raised the question of how much of the universe was currently visible from the Earth [second extract]. It also meant that we see distant objects not as they are now, but as they were at some time in the past. The fact that the speed of light, though finite, is very high was thus known to Newton when he published his major work, *Opticks*, in 1704. It is mentioned as part of Thomson's eulogy of Newton written on the latter's death in the 1720s [third extract]. The main purpose of Newton's book was to initiate the modern study of 'physical optics' – the examination of the nature and physical properties of light. The most important result was the demonstration that sunlight, when passed through a glass prism, splits up into a spectrum of colours going from red to violet: a discovery described in the final two lines of the third extract. It was one of Newton's most famous discoveries, commemorated, as Wordsworth wrote, by the statue in Trinity College, Cambridge, 'of Newton with his prism and silent face'.

Literary texts are dotted with references to the rainbow [fourth extract]. It had been recognised long before Newton that rainbows were caused by the interaction of sunlight with raindrops in the atmosphere. Newton's work now explained why this led to a bow made up of the primary colours. In direct opposition to Thomson, the Romantics of the nineteenth century felt that Newton's work had destroyed the magic of the rainbow [fifth extract]. The 'he' referred to in this extract was Charles Lamb. Benjamin Haydon, the painter, had invited Lamb and Keats to a dinner along with others (including Wordsworth). They all ended by drinking a toast to 'To Newton's health and confusion to mathematics'. The group accepted Newton's theory of light, but disliked it. Their contemporary, Johann Goethe, believed that Newton had got it wrong. In his book *On the theory of colours*, published in German in 1810, he proposed a new way of thinking about light [sixth extract]. In essence, his book was about the perception of colour, rather than the physical nature of light. His conclusions were therefore rejected by physicists, but proved to be of interest to painters and philosophers. In Britain, both Turner and the Pre-Raphaelites studied what he had written. One problem with studying the properties of light was that the only intense source of light was the Sun, which might well be hidden behind clouds During the nineteenth century, bright artificial sources were developed. An important one was the electric arc. This depended on an electric discharge between two carbon electrodes. Invented by Humphry Davy early in the nineteenth century, the arc came into widespread use for public lighting in the 1870s [seventh extract]. In science, it was particularly used for spectroscopic studies.

Early in the nineteenth century, the German scientist, Joseph von Fraunhofer, made a detailed study of the solar spectrum, and found it contained a number of dark spaces. These came to called 'Fraunhofer lines'. Later in the century, it was realised that these lines were places in the spectrum where chemical elements in the solar atmosphere were absorbing light. The same elements when heated in the laboratory gave out bright lines at the same places (i.e. at the same wavelengths). Hence a comparison of laboratory and solar spectrum could provide knowledge of the Sun's chemical composition. Consequently, in the latter part of the nineteenth century much effort was put into the study of spectra, for both terrestrial and astronomical purposes. Maxwell's poem mentions some of the most prominent lines in the spectrum: the two yellow lines of sodium (labelled D by Fraunhofer), the green b-lines of magnesium, and the thallium lines, also in the green [eighth extract] One puzzling discovery was that some of the spectral lines produced in an electric arc could differ from those usually given by a particular element. It was gradually realised that atoms in the arcs were being ionised (having electrons removed), at which point they produced a different kind of spectrum. A proper understanding of spectral lines awaited Niels Bohr's picture of the atom published in the early twentieth century. He suggested that electrons could only circulate round the nuclei of atoms at certain fixed distances.
Spectral lines resulted when an electron jumped from one permitted orbit to another [final extract].

The tendrils of my soul are twined
With thine, though many a mile apart.
And thine in close coiled circuits wind
Around the needle of my heart.
Constant as Daniel, strong as Grove.
Ebullient throughout its depths like Smee,
My heart puts forth its tide of love,
And all its circuits close in thee.
[James Clerk Maxwell *Valentine by a Telegraph Clerk*]

Lone and discarded one! Divorced by fate
Far from thy wished-for fellows - whither art flown
Where lingerest thou in thy bereaved estate,
Like some lost star, or buried meteor stone?
[J.J. Sylvester *To a missing member of a family group of terms in an algebraical formula*]

I know this little thing
A myriad men will save.
O Death, where is thy sting?
The victory, O Grave?
[Ronald Ross *Success in malaria research*]

Now twenty years ago
This day we found the thing;
With science and with skill
We found: then came the sting -
[Ronald Ross *The anniversary*]

"Better leave it alone," Towton counselled unnecessarily. "Wooton'll want to go over it for fingerprints, likely." Constable Towton had a profound contempt for the more scientific methods in criminal-hunting.
[J.J. Connington *In whose dim shadow*]

Weary of plodding science, where the vision
Must for achievement clothe itself in clay,
Where there is no completeness past revision,
But fact on fact for ever and a day.
[Julian Huxley *To a dancer*]

I wish I had the voice of Homer
To sing of rectal carcinoma,
Which kills a lot more chaps, in fact,
Than were bumped off when Troy was sacked.
[J B S Haldane *Cancer's a Funny Thing*]

I dashed to the library at the first opportunity; I refer to the venerable library of the University of Turin's Chemical Institute, at that time, like Mecca, impenetrable to infidels and even hard to penetrate for such faithful as I. One had to think that the administration followed the wise principle according to which it is good to discourage the arts and the sciences: only someone impelled by absolute necessity, or by an overwhelming passion, would willingly subject himself to the trials of abnegation that were demanded of him in order to consult the volumes.
[Primo Levi *The periodic table*]

Literary scientists DISCUSSION

Plenty of scientists have tried to write poetry or fiction. A few, such as Humphry Davy [the subject of another entry], have been well-recognised for their literary contributions. I look here at some of the others who are still occasionally remembered for their writings. The most famous scientist in this category is James

Clerk Maxwell, one of the greatest physicists of all time. Maxwell's poems were not restricted to science topics - which is true also of the other authors quoted here - but he is best known for his humorous science verse. In Maxwell's lifetime, the electric telegraph transformed communication, and scientists were deeply involved in its development. The telegraphist sent messages along a wire by using a device that could interrupt the current in the electrical circuit. In the early days, the recipient recorded this by observing the resultant motion of a needle placed at the centre of an electrical coil. Maxwell's poem goes on to note some of the batteries used to provide electrical current for telegraphy - the Daniel, Grove and Smee cells - all named after their inventors.

Sylvester, a contemporary of Maxwell, was a noted mathematician and eccentric. He held chairs at various British and American universities. The moves were due partly to employment problems caused by his Jewish background and partly because of his own ebullient nature. He was devoted to poetry, both writing his own verse, and translating poems from a number of languages into English. Matthew Arnold encouraged him to write a book on the *The laws of verse*, of which he was extremely proud. The poem quoted here was given - somewhat to the surprise of the audience - during his inaugural lecture as Professor of Geometry at Oxford.

Biologists, too, wrote poetry. Ronald Ross was born in India into a family with long-standing connections to the country. He qualified as a doctor in London, then returned to India. During the 1890s, he carried out a series of investigations into malaria and the role of mosquitoes in its transmission. The success of this work led to him receiving a Nobel Prize in 1902. The first Ross extract is from the triumphant poem he wrote when he was finally sure he had worked out the link. To his annoyance, however, his recommendations for controlling malaria were implemented only very slowly, which led to him writing the second extract. Ross wrote novels, as well as poems, though he is best remembered for the latter. He was, for a time, the president of the Poetry Society. As the two extracts quoted here may suggest, he was also somewhat self-centred.

The link between academics and detective fiction has often been noted, but relatively few of the academic authors have been scientists. J.J. Connington was a pseudonym adopted by Alfred Walter Stewart, a member of the Chemistry Department at Glasgow University. He specialised in organic chemistry, but was also interested in radioactivity. Connington became one of the leading names in detective fiction between the Wars, but - as the fifth extract may suggest - his novels were not particularly noted for their inclusion of science. Julian Huxley was a grandson of T.H. Huxley, and was one of the leaders between the Wars in the attempt to provide a new synthesis of ideas on evolution and genetics. He was also from early on interested in the popularisation of science. H.G. Wells persuaded him to become a co-author of his best-selling book *The science of life*. After the Second World War, Huxley became Director-General of the newly formed UNESCO. He is best remembered for his prose writings, but he also wrote occasional poetry. As the extract quoted here indicates, he recognised that science has its tedious aspects, as well as its exciting ones.

His contemporary, J.B.S. Haldane, held posts in genetics at Oxford and Cambridge; then moved to London in the 1930s where he became a leading expert in biometrics. He was widely known for his contributions to magazines and newspapers, not least the *Daily Worker* (he was a Communist for some years), and he wrote a very popular children's book. However, his best known piece of writing is probably the poem quoted here. At the end of his life, Haldane moved to India and became an Indian citizen. He died there of the rectal carcinoma he mentions in the seventh extract - though not before poking fun at it in verse.

Finally, Primo Levi made his reputation as a writer after the Second World War. He was an Italian Jew, who took a degree in chemistry at Turin University, graduating in 1941. After Italy signed an armistice with the Allies, Levi was captured by the Germans and sent to Auschwitz. He was liberated from there by Soviet troops and post-war worked in the chemical industry in Italy. It was during this period that he started to gain a reputation as a writer. *The periodic table* is a series of autobiographical essays, each essay woven round a particular element in the periodic table [devised by Medeleev in the nineteenth century]. In 2006, it was selected as the best science book ever in a competition held by the Royal Institution.

Mars his true moving, even as in the heavens
So in the earth, to this day is not known
[William Shakespeare *Henry the Sixth (Part I)*]

... two lesser stars, or *Satellites*, which revolve about *Mars*; whereof the innermost is distant from the Center of the primary Planet exactly three of his Diameters, and the outermost five; the former revolves in the Space of ten Hours, and the latter in Twenty-one and a Half; so that the Squares of their periodical Times, are very near in the same Proportion with the Cubes of their Distance from the Center of *Mars*; which evidently shows them to be governed by the same Law of Gravitation, that influences the other heavenly Bodies.
[Jonathan Swift *Gulliver's travels (Voyage to Laputa)*]

And as with optic glasses her keen eyes
Pierced thro' the mystic dome .....
She saw the snowy poles of moonless Mars
[Alfred Tennyson *The Palace of Art*]

The planet Mars, I scarcely need remind the reader, revolves about the sun at a mean distance of 140,000,000 miles, and the light and heat it receives from the sun is barely half of that received by this world. It must be, if the nebular hypothesis has any truth, older than our world; and long before this earth ceased to be molten, life upon its surface must have begun its course. The fact that it is scarcely one seventh of the volume of the earth must have accelerated its cooling to the temperature at which life could begin.... Men like Schiaparelli watched the red planet ... but failed to interpret the fluctuating appearances of the markings they mapped so well..... Lavelle of Java set the wires of the astronomical exchange palpitating with the amazing intelligence of a huge outbreak of incandescent gas upon the planet..... I might not have heard of the eruption at all had I not met Ogilvy, the well-known astronomer, at Ottershaw. He was immensely excited at the news, and in the excess of his feelings invited me up to take a turn with him that night in a scrutiny of the red planet.
[H.G. Wells *The war of the worlds*]

In the shadows of the forest that flanks the crimson plain by the side of the Lost Sea of Korus in the Valley Dor, beneath the hurtling moons of Mars, speeding their meteoric way close above the bosom of the dying planet, I crept stealthily along the trail.
[Edgar Rice Burroughs *The Warlord of Mars*]

... from his observatory he spies
the Martian tribes following canals
across the thirsty Martian
earth to their extinction. [Patrick McGuinness *The canals of Mars*]

Red on the south horizon, brighter than
For fifteen years, the little planet glows
.... and by the next close opposition
Observers in the exosphere
Should see it many times as clear,
And by the next one yet, match touch with vision.
[Robert Conquest *For the 1956 opposition of Mars*]

Mars   DISCUSSION

Back in the days when the Earth was thought to be the centre of the solar system, it was supposed that the paths of the other planets round it could be explained in terms of some combination of circles. Mars proved particularly difficult to pin down: it always strayed from its expected path. So Shakespeare comments [first extract] that its ways in the heavens were as uncertain as its astrological influence on Earth. In fact, these deviations of Mars provided the clue for Kepler's breakthrough in the understanding of planetary orbits. He believed the solar system to be heliocentric, and, after much labour, managed to show, early in the seventeenth century, that Mars actually followed an elliptical orbit round the Sun.

Some years later, Kepler demonstrated that there was a simple relationship between the distance of a planet from the Sun and its orbital period. His results were empirical - based on the observations he had - but later in the seventeenth century, Newton showed that they could be derived theoretically from his own ideas on motion and gravitation. One consequence was that the orbits of satellites going round planets could be explained in the same terms as the orbits of planets going round the Sun. This is precisely what Swift says [second extract]. The oddity here is that no moons had been observed round Mars when Swift was writing in the eighteenth century. There was, however, a widespread belief in the seventeenth and eighteenth centuries that Mars ought to have two moons. The argument was that Venus had no moons, the Earth had one, and Jupiter had four. This progression of numbers in order away from the Sun suggested that Mars ought to have two moons. (Later in the century, Voltaire, perhaps influenced by Swift, also suggested that Mars had two moons.) By the nineteenth century, this belief had faded away, as the third extract (published in the first half of that century) indicates.

Interest in the nineteenth century concentrated more on the features visible on Mars. Of these, the most obvious were the bright polar caps [third extract]. These varied in size with the Martian season, leading William Herschel to suggest that they were similar to the polar caps on Earth. Various attempts were subsequently made to pin down the main features of the Martian surface. This led to a detailed map produced by the Italian astronomer, Schiaparelli, in 1877. The map included a number of linear features which he labelled *canali* (ie. channels). Lowell in the USA took up the study of these channels, and eventually concluded that they actually were artificial canals with surrounding irrigated areas. As Wells [fourth extract] rehearses, there were grounds at that time for supposing Mars was an older and more evolved planet than the Earth. This gives the scenario for Wells' novel. Water was running out on Mars - hence the canals - and the planet was dying. So the Martian inhabitants decide to invade the much luckier Earth.

Since Wells' story was set on Earth, it says little about Mars, itself. Others were less reticent. Twenty years later, Edgar Rice Burroughs published his first Martian science fiction. As extract five indicates, he accepted the same picture of Mars as a dying world, along with the interpretation of the reddish areas on Mars - whence its name 'the red planet' - as arid plains. By this time, the two moons of Mars had been detected (by Asaph Hall in 1877). The inner moon of the two was found to have an orbit not too different from what Swift had guessed 150 years before. The sixth extract, though published in 2004, accurately reflects the kind of picture of Mars that Wells and Burroughs had in mind.

The best observations of Mars are obviously made when it is closest to the Earth. That is when Mars is in opposition (ie. on the other side of the Earth from the Sun). But because Mars has an elongated orbit, not all oppositions are equally good. Oppositions where Mars is as close to the Earth as possible occur every 15 years or so. One of these close oppositions is celebrated in the seventh extract. This was the last such opposition before the dawn of the space age, and Conquest speculates what will be happening at the next two (in 1971 and 1988). By the first of these, observations were ahead of Conquest's schedule. Instead of observing from the Earth's exosphere, the US had already sent probes to fly past Mars and take pictures of its surface. They showed a barren, crater-filled surface with no sign of canals or similar features. The next stage actually started in 1971 with the dispatch of a probe to orbit Mars and scan more of the surface than was possible with flypasts. This showed the surface to be a good deal more interesting than initially thought (though still no canals). As for the 1988 opposition, it depends on what Conquest meant by 'touch'. Human beings, of course, have not yet visited Mars, but probes were already landing on the surface and examining its nature in the mid-1970s.

Here, look at medicine. Big wigs, gold-headed canes, Latin prescriptions, shops full of abominations, recipes a yard long, "curing" patients by drugging as sailors bring a wind by whistling, selling lies at a guinea apiece,- a routine, in short, of giving unfortunate sick people a mess of things either too odious to swallow or too acrid to hold, or, if that were possible, both at once ..... A scheming drug-vender, (inventive genius,) an utterly untrustworthy and incompetent observer, (profound searcher of Nature,) a shallow dabbler in erudition, (sagacious scholar,) started the monstrous fiction (founded the immortal system) of Homoeopathy. I am very fair, you see,- you can help yourself to either of these sets of phrases.
[Oliver Wendell Holmes *The professor at the breakfast table*]

..... about 1829 the dark territories of Pathology were a fine America for a spirited young adventurer. Lydgate was ambitious above all to contribute towards enlarging the scientific, rational basis of his profession. The more he became interested in special questions of disease, such as the nature of fever or fevers, the more keenly he felt the need for that fundamental knowledge of structure which just at the beginning of the century had been illuminated by the brief and glorious career of Bichat, who died when he was only one-and-thirty, but, like another Alexander, left a realm large enough for many heirs.
[George Eliot *Middlemarch*]

Straighten your quilts, and be decent!
Here's the Professor.
In he comes first
With the bright look we know,
From the broad, white brows the kind eyes
Soothing yet nerving you. Here at his elbow,
White-capped, white-aproned, the Nurse, Towel on arm and her inkstand
Fretful with quills. [W.E. Henley *In hospital*]

These things I said to the Dr while Neil Munro stood in front of a Röntgen machine and on the screen behind we contemplated his backbone and his ribs. The rest of that promising youth was too diaphanous to be visible.[Edward Garnett *Letters from Joseph Conrad*]

 When he had obtained a satisfactory toxin, Martin began his effort to find an antitoxin. He made vast experiments with no results. Sometimes he was certain that he had something, but when he rechecked his experiments he was bleakly certain that he hadn't. Once he rushed into Gottlieb's laboratory with the announcement of the antitoxin, whereupon with affection and several discomforting questions and the present of a box of real Egyptian cigarettes, Gottlieb showed him that he had not considered certain dilutions. With all his amateurish fumbling, Martin had one characteristic without which there can be no science: a wide-ranging, sniffing, snuffling, undignified, unselfdramatizing curiosity, and it drove him on.
[Sinclair Lewis *Arrowsmith*]

It is our belief that every one of these processes following the initial questioning and examination can be carried out more reliably by electronic machinery. It is possible to programme a modern computer with far more facts about the relation of symptoms to disease than any man can possibly carry in his head. ..... The machine cannot forget, it cannot grow tired, it cannot have an off day. If you feed into it the symptoms and signs of a particular patient, it will notify you of the possible alternative causes of these
[John Rowan Wilson *Hall of mirrors*]

Health care has become the new name for medicine and health care delivery is what doctors now do ..... Meanwhile, we are paying too little attention, and respect, to the built-in durability and sheer power of the human organism ... .the absolute marvel of good health that is the real lot of most of us, most of the time.
[Lewis Thomas *The lives of a cell*]

In the eighteenth century, medical activity was divided between physicians, often university trained, and surgeons and pharmacists, usually trained on the job. Physicians were limited in what they could recommend – mostly using either drugs or physical approaches, such as letting blood. Things began to change in the nineteenth century. Oliver Wendell Holmes Senior (referred to thus because his son, of the same name, became a famous American judge) was trained as a physician at Harvard in the 1830s and later returned to teach there. He pressed for changes in medical work, especially for hygiene at childbirth to offset the chance of catching puerperal fever. He was also a noted poet and writer. The first extract reveals his disgust that much medical assistance was still stuck in the eighteenth century. Homeopathy had originated in Germany at the end of that century. Though it gained some strong supporters, Holmes was opposed to it on the grounds that it lacked scientific credibility – a belief which continues to be held today.

The story of *Middlemarch* is set at about the time Holmes was training at Harvard [second extract]. Bichat was a French medical researcher at the end of the eighteenth century. He pioneered the study of histology in medicine – that is the study of human cells and tissues. He believed that diseases attacked specific tissues, rather than whole organs. After his death in 1802, his ideas were followed up by others. George Eliot, writing in the 1870s, presented her forward-looking doctor, Lydgate, as one of Bichat's disciples. In terms of medical practice, one of the great advances of the nineteenth century was the introduction of antiseptic surgery. In 1867, Joseph Lister suggested that carbolic acid [phenol] should be used for the sterilisation of wounds and surgical instruments in operations. A few years later, W.E. Henley spent three years in hospital under Lister's care, during which time he wrote his poem *In hospital* [third extract]. Henley suffered from tuberculosis. Not long before his stay in hospital, he had had his left leg amputated below the knee (so providing his friend, Robert Louis Stevenson, with the idea for Long John Silver). It was thought that his right leg might also need to be amputated, but, under Lister's care, he managed to retain it.

One unexpected medical spin-off from scientific research came right at the end of the nineteenth century. In 1895, Wilhelm Röntgen in Germany discovered a new type of radiation produced by an electrical discharge through a vacuum tube. He found that the radiation, which he called X-rays, could pass through human flesh revealing the bones within. His discovery was put to use in medical practice remarkably quickly: as is reflected in this note from 1898 by Conrad [fourth extract]. (Neil Munro was a Scottish journalist who was on friendly terms with a number of contemporary writers.) Sinclair Lewis's novel explores the rise of a medical researcher from his beginnings in a small Midwest town to the directorship of a research centre [fifth extract]. Lewis was advised on the medical background by a microbiologist, Paul de Kruif (who received a share of the royalties). Lewis's hero, Martin Arrowsmith, learns about research under his mentor, Max Gottlieb, at a research centre in New York. Some of Gottlieb's characteristics derive from two contemporary researchers at the Rockefeller Institute for Medical Research in New York with whom de Kruif was acquainted - Novy and Loeb. The latter died shortly before the novel was published. Arrowsmith, himself, shares some characteristics with the French-Canadian microbiologist, Felix d'Herelle, who did pioneering work on bacteriophages, and appears in the novel as a leading competitor in Arrowsmith's area of research.

John Rowan Wilson [his actual name was John Robinson Wilson] was a surgeon who became medical director of a pharmaceutical firm, then went into medical publishing, and ultimately turned to writing fiction. *Hall of mirrors* appeared in 1966. One of his protagonists is an early advocate of the use of computers in diagnosis [sixth extract]. At the time, it was strongly disputed that this would be either appropriate or useful. Some fifty years later, computers are, indeed, being used in this way. It has, for example, been claimed that IBM's *Watson* computer can diagnose lung cancer successfully in 90 per cent of referrals, as compared with 50 per cent for human doctors.

A recurrent theme in medical writing is the need to limit the amount of intervention by doctors to what is really necessary. Thomas's comments [final extract] simply reflect what Hippocrates – often called the father of medicine – said 25 centuries ago: 'Natural forces within us are the true healers of disease.'

It turns to sulphur, or to quicksilver,
Who are the parents of all other metals.
[Ben Jonson *The alchemist*]

The Metals, lustrous Monarchs of the Cave,
Are ductile and conductive and opaque
Because each Atom generously gave
Its own Electrons to a mutual Stake
[John Updike *The dance of the solids*]

Oh metals metals, why are you always hanging about? Is it not enough that you hold men's wrists? Is it not enough that we let you in our mouths? …..
And men love you. Perhaps it is because you soften so often.
You did, it is true, pour into anything men asked you to. It has always proved you to be somewhat softer than you really are.
[Russell Edson *Metals Metals*]

Gold, *n.* A yellow metal greatly prized for its convenience in the various kinds of robbery known as trade. ….. Gold is the heaviest of all the metals except platinum.
[Ambrose Bierce *The devil's dictionary*]

Gold is for the mistress - silver for the maid -
Copper for the craftsman cunning at his trade."
"Good!" said the Baron, sitting in his hall,
"But Iron - Cold Iron - is master of them all."
[Rudyard Kipling *Cold iron*]

The mercury sank in the mouth of the dying day.
What instruments we have agree [W,H, Auden *In memory of W.B. Yeats*]

Lead, *n.* A heavy blue-grey metal much used in giving stability to light lovers.
[Ambrose Bierce *The devil's dictionary*]

Captain Flint had hauled up a chair and was sitting at the table. 'Let's have a look at that test-tube,' he said. ….. Drop by drop he let the ammonia trickle down the tube. There was some more fizzing. The clear liquid clouded thickly and then turned a brilliant blue. 'There you are', said Captain Flint, 'Copper'.
[Arthur Ransome *Pigeon post*]

Nothing of the generous good nature of tin, Jove's metal, survives in its chloride (besides, chlorides in general are rabble, for the most part ignoble by-products, hygroscopic, not good for much: with the single exception of common salt, which is a completely different matter).
[Primo Levi *The Periodic Table*]

You all probably know that the ochreous stain, which, perhaps, is often thought to spoil the basin of your spring, is iron in a state of rust ….. the ochreous dust which we so much despise is, in fact, just so much nobler than pure iron, in so far as it is *iron and the air*. Nobler, and more useful - for, indeed, as I shall be able to show you presently - the main service of this metal, and of all other metals, to us, is not in making knives, and scissors, and pokers, and pans, but in making the ground we feed from, and nearly all the substances first needful to our existence. For these are all nothing but metals and oxygen.
[John Ruskin *In praise of rust*]

In alchemy [the forerunner of chemistry], the metals were believed to form from various combinations of sulphur and mercury [first extract]. Gold, for example, required red sulphur plus mercury; silver came from an admixture of white sulphur and mercury; and so on. (The name *quicksilver* for mercury comes from an old English word and means 'living silver' – a reference to the fact that it is the only metal which is liquid at room temperatures.) Contrast with this, the modern understanding provided in the second extract. Metallic elements are ones with atoms which can readily lose electrons from their outer shell. As a result, they are good conductors of electricity and heat. Other spin-offs are that they are opaque and shiny in appearance, and often ductile (capable of being drawn out into wires). Edson adds some other properties [third extract]. Metals are malleable (they can be pressed or hammered into different shapes), and so can be used to produce a variety of objects - from bracelets to tooth fillings. Alternatively, the metal can be shaped by heating until it is a liquid and then pouring into a mould. So, although many metals are noted for their hardness, yet they can always be made soft enough to shape.

Knowledge of metals dates back to pre-history, since some of them can be found in their native state. Gold may have been the first to be discovered, but silver, copper, tin, and iron (from meteorites) also date back to antiquity. Bierce is right that platinum is heavier than gold [fourth extract], but we know now that osmium is denser still. Leaving meteorites aside, iron exists on Earth as ores in combination with other elements. It was found during the second millennium BC that iron ores could be smelted to give the metal itself. This discovery led on to the Iron Age with its development of weapons made of iron. Such weapons were to dominate fighting over next two millennia. Kipling lists succinctly the way popular metals were used in the Middle Ages [fifth extract]. Like iron, mercury exists naturally in combination. Its commonest ore is the sulphide, which is bright red. (It was used as a pigment by Palaeolithic cave painters.) Mercury can easily be obtained from this by heating. The metal has long been used in instruments such as barometers and thermometers, since it has appropriate properties: for example, it does not stick to glass [sixth extract]. As with mercury, the commonest lead ore is the sulphide, and it, too, has been used as a pigment - in this case, black in colour – from antiquity. The Romans made great use of lead, some of it obtained from Britain, for such things as water pipes. Bierce's description of the metal as 'blue-grey' is only true when it is freshly cut, for it tarnishes rapidly in air [seventh extract]. Its heaviness is proverbial and receives a mention in the Old Testament.

Although some metal ores are easily identified by eye, it is customary to confirm the identification by chemical tests. In this Arthur Ransome story, the children believe they have found gold in the Lake District [eighth extract]. Captain Flint, however, shows that what they have found is actually a copper ore. The test he uses involves dissolving the copper in hydrochloric acid and adding ammonia: in the presence of air, this produces a deep-blue compound. (Gold has been found in the Lake District, but copper is much commoner there.) Copper metal has been obtained via the smelting of its ores since around 5000 BC. It was subsequently found that the addition of a suitable amount of tin produced a harder alloy that could be used for tools and weapons. So, around 3000 BC, the Stone Age gave way to the Bronze Age. In antiquity the gods were associated with particular metals. Venus, for example, was associated with copper, while Jupiter was associated with tin. The link with Venus may be explained by the use of copper for mirrors. The link between Jupiter and tin is less clear, though Levi presumably has bronze as one of the things in mind when he praises the metal [ninth extract]. The chloride, in fact, is not totally worthless, for it has been used as a mordant in dyeing. A hygroscopic substance is one that readily absorbs moisture from the atmosphere. Thus copper chloride is brown, but it rapidly absorbs water from the atmosphere and turns blue. Salt [sodium chloride] is hygroscopic, but only when the humidity is high.

As Ruskin says, rust is just another form of iron: it is iron oxide formed from the interaction of iron with moist air [final extract]. Research since Ruskin's time has underlined the importance he assigns to iron in combination for living processes. Iron meteorites are found in pristine state because their iron is mixed with nickel, and the alloy, unlike pure iron, is not susceptible to rust.

These burning fits but meteors bee,
Whose matter in thee is soone spent.
Thy beauty and all parts, which are thee,
Are unchangeable firmament.
[John Donne *A feaver*]

Oft in this season, silent from the north
A blaze of meteors shoots: ensweeping first
The lower skies, they all at once converge
High to the crown of heaven, and all at once
Relapsing quick, as quickly reascend
[James Thomson *The seasons: Autumn*]

'Tis said, she first was changed into a vapour .....
And then into a meteor, such as caper
On hill-tops when the moon is in a fit
[Percy B. Shelley *The witch of Atlas*]

As often thro' the purple night,
Below the starry clusters bright,
Some bearded meteor, trailing light,
Moves over still Shalott.
[Alfred Tennyson *The Lady of Shalott*]

Great fall of stars, identified with Biela's comet. They radiated from Perseus or Andromeda and in falling, at least I noticed it of those falling at all southwards, took a pitch to the left halfway through their flight. The kitchen boys came running with a great to do to say something red hot had struck the meatsafe over the scullery door with a great noise and falling into the yard gone into several pieces. No authentic fragment was found, but Br. Hostage saw marks of burning on the safe and the slightest of dints as if made by a soft body, so that if anything fell it was probably a body of gas, Fr. Perry thought.
[Gerard Manley Hopkins Journal: 1872]

I find nothing surprising in the raining of stones in France, nor yet had they been mill-stones. ….. The reason is that the exuberant imagination of a Frenchman gives him a greater facility of writing, and runs away with his judgment unless he has a good stock of it. It even creates facts for him which never happened, and he tells them with good faith.
[Thomas Jefferson Letter to Andrew Ellicott, 1803]

[A labourer] brought him a stone of the colour of lead, weighing ten pounds, and irregular in its figure, which stone the labourer had found ….. for after the two reports and the rumbling sound, he heard some heavy body fall near him, and found this stone sunk into the ground, still warm, and the ground freshly moved
[Robert Southey Letters written during a short residence in Spain and Portugal]

At seven o'clock a hollow noise was heard, and a thud so tremendous was felt that the island shook on its base. A few moments later one of the natives ran to the house occupied by Monsieur de Schnack. He was the bearer of the great tidings. The bolide had descended on the north-west point of the island of Upernivik.
[Jules Verne The chase of the golden meteor]

Meteors   DISCUSSION

The meaning of the word meteor has evolved over time. In the medieval world picture, the universe was divided into two parts – the Earth and the heavens. These two parts had different characteristics. In particular, change could only occur on Earth: the heavens were changeless. It followed that any sudden change in the sky must be an atmospheric phenomenon. Such changes were labelled 'meteors', so their study was 'meteorology'. Donne sees an analogy between a temporary meteor in the sky and a temporary illness on Earth [first extract].

By the eighteenth century, when Thomson was writing, the division between the Earth and the heavens no longer existed. However, the word meteor was still used in a variety of contexts. For example, there was the well-known phenomenon of the 'Northern lights'. These were best seen at higher latitudes, though sometimes mentioned by classical writers further south. They received their official name, aurora borealis, in the seventeenth century. At the beginning of the following century, Edmund Halley explained them as the result of the impact of particles, influenced by the Earth's magnetic field, on the atmosphere. This is the explanation – with subatomic particles substituted for particles - accepted today, but, at the time, there were several competing theories. For example, one suggested that the lights were due to the emission of gas from the Earth's surface. Thomson, like his contemporaries, still thought of them as meteors [second extract]. Shelley's meteor [third extract] does relate to gas emission from the Earth's surface. The will o' the wisp phenomenon – lights seen at night hovering over the surface of boggy areas – had long featured in folk stories. Around the time that Shelley was born, it was proposed – for example, by Joseph Priestley – that they were due to some kind of electrical discharge through marsh gas [methane]. This seems to be the picture that Shelley had in mind. (Unlike the aurorae, their exact explanation remains unclear.)

During the nineteenth century, the word meteor came to be restricted to what was popularly called a 'shooting star' - much as it is today. It meant a piece of extra-terrestrial material that hit the Earth's atmosphere and burnt up, leaving a temporary tail behind it. This was the picture Tennyson had in mind [fourth extract]. Most meteors are solitary and come from any direction. Occasionally, a whole shower of meteors appears, in which case they appear to radiate out from a particular part of the sky. It was realised in the 1860s that such meteor showers occurred when the Earth passed through the debris left behind by a comet. Biela's comet (named after the person who calculated its orbit) disintegrated around the middle of the nineteenth century. However, in 1872 the Earth passed through the former comet's orbit and a remarkable meteor shower ensued. This is the shower that Hopkins observed [fifth extract]: it actually came from the direction of the constellation Andromeda. (The apparent change of direction he mentions was probably an optical illusion.) It has been speculated that some fragments could have reached the Earth's surface, and perhaps Hopkins witnessed such an event. The Father Perry he mentions was in charge of the observatory at Stonyhurst School.

The idea that material from outer space could reach the Earth's surface was only accepted during the nineteenth century. A detailed investigation by the leading French scientists of a fall of material in 1803 led eventually to a general acceptance that such events could occur. Initially, however, the idea encountered widespread scepticism – as reflected by the comments of Thomas Jefferson, then President of the United States [sixth extract]. Southey presents an account of such a fall that happened a few years later [seventh extract]. Such falls are often accompanied by sound and a bright trail in the sky (less obvious in a daytime fall like this). The retrieved material often has a dark skin, caused by the heat from friction with the atmosphere. This air friction also slows the object down so that it usually only produces a small crater on impact with the Earth.

Verne's fictional account emphasizes the same features. It also underlines the variety of terms that have been used to describe the incoming material. Bright meteor trails are called fireballs, and the brightest of all can be termed bolides. Most meteors are small particles, often only the size of grains of sand, which burn up totally in the atmosphere. Bolides are usually caused by much larger lumps of material. If some of this lump survives passage through the atmosphere, it is called a meteorite. In times past, the definitions were vaguer, which is why Verne's book refers to the golden 'meteor', rather than – as it would be nowadays – the golden 'meteorite'.

2 Herald. Certain and sure news.
1 Herald. Of a new world.
2 Herald. And new creatures in that world.
1 Herald. In the orb of the moon.
2 Herald. Which is now found to be an earth inhabited.
1 Herald. With navigable seas and rivers.
2 Herald. Variety of nations, polities, laws.
1 Herald. With havens in 't, castles, and port-towns.....
2 Herald. But differing from ours.
[Ben Jonson *News from the New World Discovered in the Moon*]

The discovery of a new world; or, a discourse tending to prove, that (it is probable) there may be another habitable world in the Moon ....
The spots represent the sea and the brighter parts the land. .... we may guess in the general that there are some inhabitants in that planet: for why else did providence furnish that place with all such conveniences of habitation as have been above declared?
[John Wilkins *Book title in italics above*]

Having thus finished this discourse, I chanced upon a late fancy to this purpose, under the feigned name of Domingo Gonsales, written by a late reverend and learned bishop: in which .... there is delivered a very pleasant and well-contrived fancy concerning a voyage to this other world. He supposeth that there is a natural and usual passage for many creatures betwixt our earth and this planet. Thus he supposeth the swallows, cuckoos, nightingales, with divers other fowl, which are with us only half the year, to fly up thither, when they go from us.
[John Wilkins *op.cit.*]

It is the opinion of Kepler, that as soon as the art of flying is found out, some of their nation will make one of the first colonies that shall transplant into that other world. I suppose that his appropriating this pre-eminence to his own countrymen may arise from an over-partial affection to them. But yet this far I agree with him, that whenever that art is invented, or any other, where a man may be conveyed some twenty miles high, or thereabouts, then it is not altogether improbable that some or other may be successful in this attempt.
[John Wilkins *op.cit.*]

A learned society of late,
The glory of a foreign state,
Agreed, upon a summer's night,
To search the Moon by her own light;
To make an inventory of all
Her real estate, and personal;
And make an accurate survey
Of all her lands, and how they lay .....
And make the proper'st observations
For settling of new plantations,
If the society should incline
T' attempt so glorious a design.
[Samuel Butler *The elephant in the Moon*]

One of the great events in the history of science was Galileo's publication of the *Siderius Nuncius* [*The Starry Messenger*] in 1610. The title page elaborates: 'The Starry Messenger, revealing great, unusual, and remarkable spectacles .... as observed by Galileo Galilei, ... with the aid of a spyglass, lately invented by him, in the surface of the Moon'. Though telescopes existed before Galileo, they had not been used for a systematic study of the heavens. The book made an enormous impact. Sir Henry Wotton - best remembered for his remark, 'An ambassador is an honest gentleman sent to lie abroad for the good of his country' - was in Italy at the time representing King James I. He reported back in great excitement: 'I send herewith unto his Majesty the strangest piece of news that he hath ever yet received from any part of the world; which is the annexed book of the Mathematical Professor at Padua'. He went on to explain why it was so important: 'he hath first overthrown all former astronomy ..... and next all astrology'. Existing ideas of the universe were mainly Aristotelian, and supposed that things beyond the Earth were quite different in their nature from things on Earth. However, Galileo's observations of the Moon suggested that, on the contrary, the Moon had much in common with the Earth. Ben Jonson's masque was written ten years later, by which time it was popularly supposed that the similarity between the Earth and the Moon extended to both having life on their surfaces.

John Wilkins was that unusual thing - a moderate in the English civil war of the mid-seventeenth century. On the one hand, he married Cromwell's youngest sister, but, on the other, he was appointed Bishop of Chester by Charles II. More importantly, he played a leading role in the scientific discussions of the day, and became a founder-member of the Royal Society. When Wilkins published his discussion of the Moon at the end of the 1630s, he took a properly cautious line, but, as the first extract indicates, he also believed in the likelihood of lunar inhabitants. For him, as for others, this belief was linked to what was subsequently labelled 'the principle of plenitude'. The universe was expected to be as richly abundant as possible: if niches were available, they would be filled. It followed that, if conditions on the Moon were suitable for life, then all possible types of life would be present there, just as they were on Earth. (Since the Moon was not identical with the Earth, the life forms there might however be different.) Wilkins devoted some time to thinking of ways of getting to the Moon. As he notes, another book - actually written by Francis Godwin, Bishop of Hereford - had recently been published posthumously. It contained the idea of using a chariot towed by birds to reach the Moon. This was less far-fetched then than it sounds now. The belief that some migratory birds spent part of the year on the Moon actually persisted into the eighteenth century.

By the mid-seventeenth century, the idea of empire-building occupied the thoughts of several European powers. Thus the term 'British empire' had already been coined - perhaps by the mathematician, John Dee - in Elizabethan times. So it is hardly surprising that the new world of the Moon should be seen as being just as desirable an acquisition as the new lands being discovered on Earth. The German astronomer, Johannes Kepler, had written about visiting the Moon and what it would be like there earlier in the seventeenth century. As the final quotation reflects, contemporary discussion of a lunar world that might be colonised had reached the point where it could readily be parodied. Samuel Butler, a satirical poet of the time, found the newly formed Royal Society (set up after the restoration of Charles II in 1660) particularly amusing. The elephant in the Moon represents a satirical look at the experiments with telescopes then being performed by several Fellows of the Society. (The elephant of the title turns out to be a mouse caught in the telescope.)

'The Council appointed a special sub-committee to examine and report .... I suggest that Dr. Brine should give the views of his sub-committee.' 'Please,' said Gladwin a bit pathetically ....

The blue-chinned man stroked the blue part and said, 'Well, Mr. Chairman, my colleagues and I approached this matter purely as scientists.' He said it as though everybody else had approached it as income tax inspectors or jobbing gardeners. 'And our conclusion was that scientifically speaking it was not a sound conception. Not at all a sound conception. In fact I'll go further and say that no scientist could feel happy about many of the principles involved.'

[Nigel Balchin *The small back room*]

But the N.P.B. [National Power Board] was riven again, especially in the upper reaches, by the swarming of three factions referring to each other as *The Administrators*, *The Scientists*, and *The Engineers*. It was a sort of polarization on Two Cultures lines, with the technological culture in its wisdom seeing fit to present a divided front. Incidentally I assure you they really did refer not only to each other but to themselves by these titles. Any day of the week in the N.P.B. you could hear such remarks as:

'You can't reason with The Scientists.' (Administrator)

'We could get on better without The Administrators.' (Scientist)

'Of course I'm only an Engineer.' (Engineer)

[William Cooper *Memoirs of a new man*]

I went towards Drawbell's desk. ..... I had met him several times, in that office as well as in London. He was not an academic, and Luke and the others said, with their usual boisterous lack of respect, that he was not a scientist at all. In peacetime he had been head of another government station. .....

'Eliot,' he said at last, 'I'm not satisfied with the support that we're receiving'. ....

I knew in cold blood what was bound to happen. Even if Rudd's scheme worked (perhaps Martin was underestimating its chances), it would take years. All the scientists they wanted were working elsewhere, most of them on R.D.F., on work that would (as the Minister said) pay dividends in one year or two, not in the remote future: no one in authority could take the risk of moving them; even if the Barford result was certain, instead of uncertain, no one at that stage of the war could do much more.

[C.P. Snow *The new men*]

I said: 'There's absolutely no need for this meeting. The outcome's perfectly obvious before we ever start.'

Robert said: 'My dear Joe, when *will* you learn?'

'Learn what?'

'The way people like to run things.'

[William Cooper *Scenes from metropolitan life*]

I said, 'At what stage do you think we ought to get the firm in on this thing, sir?' I paused and then I added, 'E.P. Prendergast .....'

He glanced at me, 'Yes ..... Prendergast.' .... Prendergast would go up in a sheet of flame. He would complain to the Minister, as he had done before, that he could not carry on his work in an atmosphere of petty backbiting and vilification by minor civil servants.

The Director said, 'I doubt if Mr Prendergast would find Honey's theoretical work very convincing'.

'I'm damn sure he wouldn't,' I said. 'He'd chew him up and spit him out in no time.'

'I don't know that the time is quite ripe to inform the firm,' he said thoughtfully.

[Nevil Shute *No highway*]

Though science had played a part in the First World War, it became dominant in the Second World War. By the end of the war it was evident that scientists and engineers would be playing a much enlarged role in determining future policy. The question, especially for civil servants, was how to work with the newcomers. During the war, scientists and engineers tended to work in the background - not least because much of their work was classified. As Lord Beaverbrook (recruited by Churchill as Minister of Aircraft Production) explained in 1941: 'Now who is responsible for this work of development on which so much depends? To whom must the praise be given? To the boys in the back rooms. They do not sit in the limelight. But they are the men who do the work.' From this came their nickname - 'backroom boys'. It was popularised by Nigel Balchin's novel *The small back room* published in 1943, along with the alternative nickname of 'boffins', which was publicised post-war in Nevil Shute's novel *No highway*. (Both novels, incidentally, were turned into films.) The first extract is taken from the description of a committee meeting where an idea for a new gun is being discussed. The dissension here is between two groups of scientists and engineers - as, indeed, often happened during the war - rather than with outsiders.

The second extract comes from a post-war novel (published in 1966) when scientists and engineers had become embedded in the civil service and in policy making. The divisions Cooper describes were not restricted to the UK. Administrators everywhere liked asserting that scientists should be on tap, not on top. Equally, engineers worldwide could feel undervalued relative to scientists. Thus, in the 1960s, US space engineers were prone to claim that a successful satellite launch was called a scientific achievement, whereas an unsuccessful one was labelled an engineering failure. William Cooper [real name Harry Hoff] knew about all this at first hand as a member of the Civil Service Commission. He was a friend and colleague of C.P. Snow's. Indeed, Snow saw Cooper as a prototype 'new man': a bright, scientifically educated boy from a lower-middle-class background, purposefully at large in the corridors of the mid-twentieth-century establishment (as the *Oxford DNB* puts it). This explains both the title of Cooper's novel and the mention of the 'Two Cultures'. The latter refers to the debate initiated by Snow in the 1950s. Snow believed that the ignorance of modern science among people on the arts side was wider and more significant than the ignorance of the arts amongst scientists. The point was that senior civil servants - the administrators - traditionally came from an arts background and were ill-qualified to cope with scientific questions.

The next piece comes from Snow's own novel of 1954, which introduced his readers to the 'new men'. The Eliot in this abstract is Lewis Eliot, an academic conscripted to the wartime civil service. Martin is his younger brother, a scientist. The two major science-based developments of the Second World War were radar and the atomic bomb. The former, under the official acronym R.D.F. [Radar Direction Finder], played a significant role throughout the war. The latter only reached viable form at the end of the war. In Snow's novel, Barford was the imagined centre of the early British work on the atomic bomb, and Martin was one of the scientists working there. Next comes an excerpt from another Cooper novel. The central figure, Joe Lunn, is a scientist involved in the procurement of equipment for the Royal Air Force. His immediate boss, Robert, is a portrait of C.P. Snow.

Finally, a scene set in a real research establishment. Nevil Shute Norway was a well-respected aeronautical engineer. (He dropped his surname for his fictional writings: apparently in fear that they might affect his reputation as an engineer.) In 1931, he established his own firm, Airspeed, for designing and manufacturing aircraft. During the Second World War, he was seconded to the Admiralty to work on secret weapons. *No highway* is set immediately post-war in the Royal Aircraft Establishment at Farnborough, then a leading centre both nationally and internationally for research on matters relating to aircraft. The RAE was funded at that time by the Ministry of Supply and thus operated at the sometimes rather difficult interface between civil servants, industry and researchers.

Gin a body meet a body
Flyin' through the air.
Gin a body hit a body,
Will it fly? And where?
[James Clerk Maxwell *Rigid body sings*]

When Newton saw an apple fall, he found
In that slight startle from his contemplation -
Tis *said* (for I'll not answer above ground
For any sage's creed or calculation) -
A mode of proving that the earth was round
In a most natural whirl, called 'gravitation'
[Lord Byron *Don Juan Canto X*]

He, first of men, with awful wing pursu'd
The comet through the long elliptic curve ....
The heavens are all his own, from the wild rule
Of whirling vortices and circling spheres
To their first great simplicity restor'd.
[James Thomson *A poem sacred to the memory of Sir Isaac Newton*]

Mark where he halts on Saturn, tipt with snow,
And pleas'd surveys his theory below;
Sees the five moons alternate round him shine,
Rise by his laws, and by his laws decline
[Samuel Bowden *A poem sacred to the memory of Sir Isaac Newton*]

Like Ministers attending ev'ry Glance,
Six worlds sweep round his Throne in Mystick Dance.
He turns their Motion from its devious Course,
And bends their Orbits by Attractive Force:
His pow'r coerc'd by Laws, still leaves them free,
Directs, but not Destroys, their Liberty
[J.T. Desaguliers *The Newtonian system of the world*]

Could he, whose rules the rapid comet bind,
Describe or fix one movement of his mind!
Who saw its fires here rise and there descend,
Explain his own beginning or his end?
[Alexander Pope *Essay on man Epistle II*]

He said that new Systems of Nature were but new Fashions, which would vary in every Age; and even those who pretended to demonstrate them from Mathematical Principles, would flourish but a short Period of Time, and be out of Vogue when that was determined.
[Jonathan Swift *Gulliver's travels Part III*]

May God us keep From Single vision & Newtons sleep.
[William Blake,   *Letter to Thomas Butt, 1802*]

Isaac Newton's *Philosophiæ Naturalis Principia Mathematica* [*Mathematical Principles of Natural Philosophy*] appeared in 1687. The book (whose title is usually abbreviated to the *Principia*) provided the framework for discussions of how bodies move for well over two centuries. In it were formulated Newton's three laws of motion along with his law of universal gravitation. Despite their apparent simplicity, applying the laws to real events often proved complicated. For example, it is simple to apply the laws of motion to the collision of two snooker balls. Applying them to complex bodies is considerably more difficult - which is the point of the poem quoted in the first extract. Its author - the great Scottish physicist, Maxwell - was parodying the poem *Coming thro' the rye* by his fellow-countryman, Robert Burns.

Newton was stimulated to think about gravitation by seeing an apple fall in his garden (at least, so he told his friends). The point behind this famous story is that he interpreted the fall of the apple as indicating that all bodies attracted each other. The Earth attracted everything in the universe - the apple, the Moon, and so on - and reciprocally everything attracted the Earth. This attraction, he found, followed a very simple law. Gravitation had other consequences. For example, it also implied that a stationary Earth would be spherical in shape - as Byron reports in the second extract. (In fact, the Earth is spinning, so it is not a perfect sphere.) But the finding that most impressed his contemporaries was that the paths of comets could be predicted [third extract]. Comets, with their unexpected appearances, had always been seen as omens of danger. Now they proved to be explicable on exactly the same basis as planets. The second part of the third extract refers to the world system favoured before the *Principia* was published - that proposed by Descartes. (In his original version of the poem, Thomson was more specific: he said that Newton had rescued heaven and Earth from the 'French dreamer'.)

Samuel Bowden was an English physician. Unlike Thomson, he was not born until after Newton's death, but his celebration of Newton [extract four] is similar - even to the title. The point he is making here is that Newton's laws apply to the satellites of planets, just as much as to the planets themselves. The development of the telescope during the seventeenth century led to the discovery of five moons revolving round Saturn. (No more were found till the end of the eighteenth century.) This was a larger number than for any other planet, and yet Newton's laws could explain their motions.

John Desaguliers worked with Newton for a time, and became a leading populariser of Newtonian ideas. Like others in the eighteenth century, he extended their implications beyond the bounds of science. In this extract, he is comparing the solar system with its six planets (as then known) with limited monarchy under the Georges. The analogy between Newtonian mechanics and government by a system of checks and balances was popular in the eighteenth century. This kind of belief was a factor behind the formulation of the US Constitution.

Eulogies of Newton were countered from the start by expressions of caution. Pope admired Newton, but, as the sixth extract shows, he cautioned against excessive adulation. He, like many, selects comets as the most remarkable result of Newton's work. Swift, on the contrary, was unimpressed by contemporary science. This is most obvious in Gulliver's voyage to Laputa; but the extract here is taken from a subsequent trip to Glubbdubdrib. There Gulliver encounters a magician who allows him to discuss history with a succession of ghosts. Swift, unwittingly, actually put his finger on the reason why Newton's work survived and proved basic for subsequent developments: the emphasis in the *Principia* on mathematical principles. This allowed the making of quantitative predictions, which could then be denied or confirmed by observation and experimentation. Blake was even more opposed. He rejected eighteenth-century rationalism in general, and Newtonian mechanics in particular. Newton had 'single vision' because he was only interested in materialistic science. One of the most famous images that Blake - both a poet and an artist - drew was of Newton. It shows him sitting under water, naked on a rock. He is bent over looking at a geometrical diagram, which he is measuring with a pair of dividers. The picture reflects Blake's view of Newton as one who ignored the wonders around him as he worked. It is, perhaps, unfortunate that this image was used to provide the basis for the striking statue in the forecourt of the British Library in London.

This onion-dome holds all intricacies
Of intellect and star-struck wisdom ....
Yet night awakes it; blind lids open
Leaden to look upon the moon:
A single goggling telescopic eye
Enfolds the spheric wonder of the sky.
[Sidney Keyes *Greenwich Observatory*]

Bomb in Greenwich Park. There isn't much so far. Half-past eleven. Foggy morning. Effects of explosion felt as far as Romney Road and Park Place. Enormous hole in the ground under a tree filled with smashed roots and broken branches. All round fragments of a man's body blown to pieces. ...... No doubt a wicked attempt to blow up the Observatory, they say.
[Joseph Conrad *The secret agent*]

After one. Timeball on the ballast office is down. Dunsink time. Fascinating little book that is of Sir Robert Ball's. ..... Now that I come to think of it, that ball falls at Greenwich time. It's the clock is worked by an electric wire from Dunsink. Must go out there some first Saturday of the month. If I could get an introduction to Professor Joly or learn up something about his family
[James Joyce *Ulysses*]

"To-morrow night"--so wrote their chief--"we try
Our great new telescope, the hundred-inch......
To-morrow night! For more than twenty years,
They had thought and planned and worked. Ten years had gone,
One-fourth, or more, of man's brief working life,
Before they made those solid tons of glass,
Their hundred-inch reflector, the clear pool,
The polished flawless pool that it must be
To hold the perfect image of a star.
[Alfred Noyes *Watchers of the sky*]

Before he left the house the new astronomical project was set in train. The top of the column was to be roofed in, to form a proper observatory; and on the ground that he knew better than any one else how this was to be carried out, she requested him to give precise directions on the point, and to superintend the whole. A wooden cabin was to be erected at the foot of the tower, to provide better accommodation for casual
visitors to the observatory than the spiral staircase and lead-flat afforded.
[Thomas Hardy *Two on a tower*]

Who were they, what lonely men
Imposed on the fact of night
The fiction of constellations
Who made the dark stable
When the light was not? Now
We receive the blind codes
Of spaces beyond the span
Of our myths
[Patric Dickinson *Jodrell Bank*]

The Royal Observatory at Greenwich was built in 1675 at the command of Charles II. Its main purpose was to aid navigation at sea. (The architect, Christopher Wren, said he had built it for the observer's habitation and a little for pomp.) Subsequently, more telescopes were added to the Observatory under a number of different domes. In the 1890s, one of the telescopes was replaced by a larger instrument. The new telescope was too large for the existing hemispherical dome, so a replacement 'onion-shaped' dome was built. This is the visually distinctive dome mentioned in the first extract. The telescope was a refractor (ie. used lenses, rather than mirrors): its main lens - the 'telescopic eye' - measured 28 inches across, making it one of the larger refractors in the world. Keyes poem was written in 1944; in the following decade increasing atmospheric and light pollution led to a move of all the work and staff to a site in Sussex.

By the nineteenth century, the Royal Observatory had a worldwide reputation. In 1884, this was underlined when it was chosen as the prime meridian from which all longitudes would be calculated. Such prominence, in turn, enhanced its value as a target for terrorists, of whom many varieties resided in London during the latter part of the century. In 1894, a French anarchist blew himself up while trying to plant a bomb at the Observatory. The press report is summarised in the extract from Conrad's book. The summary is accurate enough, but Conrad has changed the dates. *The secret agent* was published in 1907, but the story was set in 1886 - several years before the actual bomb incident.

The time-ball was introduced to Greenwich in 1833. It was a sphere, five feet in diameter, which was let fall from the top of a pole at exactly one o'clock. Vessels below on the Thames could use it to check their chronometers. Dunsink Observatory, just outside Dublin, provided a similar service in Ireland by triggering the fall of a time-ball in the city, as Joyce records. However, his times are a little confused. Dublin time, some twenty-five minutes different from Greenwich time, remained standard till 1914, and it was then that the Dublin time-ball was switched to Greenwich time. Sir Robert Ball was a noted writer of popular astronomy books (and was in charge of Dunsink from 1874 to 1892). Subsequently, C.J. Joly headed the Observatory until his death in 1906, and so was there on Joyce's Bloomsday [16 June 1904].

As telescopes became larger, the need for clear skies became increasingly important. Most often this meant building an observatory near the top of a mountain. The most famous such observatory in the first half of the twentieth century was on Mt. Wilson near Los Angeles. Noyes' poem is set in 1917 when the Observatory's new 100-inch telescope was inaugurated. The hundred-inch diameter main mirror made this the largest telescope in the world, and it was, as Noyes says, a long time in the making. As an example, the molten glass forming the mirror had to be cooled slowly for a year in order to avoid any cracking.

Astronomy has always been a popular subject with amateurs. In the nineteenth century, as today, many individuals built their own personal observatories, as Hardy's hero is doing in this extract. Hardy had a continuing interest in astronomy and prior to writing this novel actually requested a visit to Greenwich Observatory to view its activities. His hero, as his astronomical expertise grows, makes visits to observatories abroad in order to observe in better viewing conditions. He is particularly taken by the Cape Observatory in South Africa - one of the leading observatories in the southern hemisphere.

Astronomers, until relatively recently, were solely concerned with observations in the visual region of the spectrum. Other parts are mainly blocked off by the atmosphere. However, it was realised in the twentieth century that there is another 'window' in the atmosphere, this time in the radio region. After the Second World War, some of the people who had been involved in the development of radar built radio telescopes to see what new things might be discovered in this region. One of the pioneering radio observatories, sited at Jodrell Bank in Cheshire (an outstation of Manchester University), was set up in 1945 by Bernard Lovell [who died in August 2012]. Radio observations, unlike visual observations, always needed ancillary apparatus to interpret the signals received - presumably the 'blind codes' to which Patric Dickinson refers.

Shall we then say that the vegetable living filament was originally different from that of each tribe of animals above described? And that the productive living filament of each of those tribes was different originally from the other? Or, as the earth and ocean were probably peopled with vegetable productions long before the existence of animals...shall we conjecture that one and the same kind of living filament is and has been the cause of all organic life? [Erasmus Darwin *Zoonomia*]

First forms minute, unseen by spheric glass,
Move on the mud, or pierce the watery mass;
These, as successive generations bloom,
New powers acquire, and larger limbs assume;
Whence countless groups of vegetation spring,
And breathing realms of fin, and feet, and wing.
[Erasmus Darwin *The Temple of Nature*]

Hence without parent by spontaneous birth
Rise the first specks of animated earth;
From Nature's womb the plant or insect swims,
And buds or breathes, with microscopic limbs.
[Erasmus Darwin *The Temple of Nature*]

Many and long were the conversations between Lord Byron and Shelley, to which I was a devout but nearly silent listener. During one of these, various philosophical doctrines were discussed, and among others the nature of the principle of life, and whether there was any probability of its ever being discovered and communicated. They talked of the experiments of Dr Darwin (I speak not of what the Doctor really did, or said that he did, but, as more to my purpose, of what was then spoken of as having been done by him) who preserved a piece of vermicelli in a glass case, till by some extraordinary means it began to move with voluntary motion.
[Mary Shelley *Frankenstein*]

It is often said that all the conditions for the first production of a living organism are now present, which could ever have been present. But if (and oh! what a big if!) we could conceive in some warm little pond, with all sorts of ammonia and phosphoric salts, light, heat, electricity, &c., present, that a protein compound was chemically formed ready to undergo still more complex changes, at the present day such matter would be instantly devoured or absorbed, which would not have been the case before living creatures were formed.
[Charles Darwin - a letter to Joseph Hooker, 1871]

In the mud of the Cambrian main
Did our earliest ancestor dive:
From a shapeless albuminous grain
We mortals our being derive
[Grant Allen *A ballade of evolution*]

Erasmus Darwin, the grandfather of Charles Darwin, was a key figure in the interaction of science and literature during the latter half of the eighteenth century. He was a leading member of the Lunar Society of Birmingham - a small, informal dining club that included such major names as Joseph Priestley, James Watt and Josiah Wedgwood - and he was also a highly regarded physician. Darwin delighted in writing lengthy poems that set out his ideas about nature. Most of the leading poets of the early nineteenth century, from Wordsworth on, read and were influenced by his writings. Yet, after his death, his poems soon fell out of fashion, for they were at odds with views on both poetry and nature in the new Romantic era. Byron's criticism was a common one. He wrote of: 'Darwins pompous chime. /That mighty master of unmeaning rhyme, /Whose gilded cymbals, more adorn'd than clear /The eye delighted but fatigued the ear'. Nevertheless, Darwin's ideas continued to be discussed.

Erasmus Darwin was one of the early evolutionists. The first quotation finds him querying whether all life, both plant or animal, might not have developed from some more basic form of life. The second extract, from a later poem, shows him assuming that this picture of the development of life is correct. (The 'spheric glass' is a microscope. Microscopes were invented in the seventeenth century and considerably improved during the eighteenth century.) This assumption naturally led him on to ask how such simple life had itself originated. In the third quotation, he gives his answer - spontaneous generation. Belief in the possibility of life appearing out of non-living material dates back to classical times. In Erasmus Darwin's time, it was still a respectable belief. One supposed example - that maggots were spontaneously generated in rotting meat - had, indeed, been disproved by the Italian, Francesco Redi, in the seventeenth century. Yet observations using the microscope seemed to provide visual evidence that minute creatures might appear out of nothing. The discussion that Mary Shelley records in the fourth extract - an important element in the subsequent genesis of her book, *Frankenstein* - was certainly a distortion of Darwin's thoughts. The likelihood is that there was a mix up with Darwin's description of vorticelli (sub-microscopic creatures) in his poem *The Temple of Nature*. However, it reflects the contemporary interest in the origin of life.

The next extract is from a letter by Erasmus Darwin's grandson, Charles Darwin. He is writing to his old friend, Joseph Hooker, the Director of the Royal Botanic Gardens at Kew. By this time, the concept of spontaneous generation as something that was still happening was on the way out. A few years before, Louis Pasteur in France had devised an experiment which seemed to show conclusively that supposed examples of spontaneous generation were actually the result of contamination. This conclusion was used by Pasteur and others to cast serious doubt on Darwin's evolutionary ideas. So Darwin was faced with the problem of explaining how, in the light of Pasteur's experiments, life could have originated. His answer was that the conditions on the early Earth were different from today's environment: so what was possible then, is not possible now. In fact, his description of a warm little pond containing the appropriate chemicals is not all that far from the explanation that is still put forward today.

Grant Allen, the author of the last extract, was a well-known science writer in the latter part of the nineteenth century. The quotation forms part of an essay, *The evolutionist at large*, published in 1881. By this time, the idea that life originated long ago, thence providing the basis for all subsequent evolution, had become well-established. Allen refers to the 'Cambrian'. During the nineteenth century, geologists classified rocks into different groups depending, in part, on what fossils they contained. 'Cambrian' derives from the Latin word, *Cambria*, denoting Wales, where the group of rocks concerned were especially prominent. The different groups were seen as coming from different periods of time, and the Cambrian rocks were the earliest to contain fossils. So it was supposed that these rocks contained the origins of life, which explains Allen's reference to them. (Nowadays, this attribute is assigned to even older rocks - the Precambrian.)

Did I see it go by,
That Millikan mote?
Well, I said that I did.
I made a good try.
But I'm no one to quote.
If I have a defect
It's a wish to comply
And see as I'm bid.
[Robert Frost *A wish to comply*]

Now Rutherford had shown that ordinary atoms were not indestructible. By knocking out a hydrogen nucleus (later called a proton) from the nucleus of nitrogen he had converted it into another element, oxygen.
[C.P. Snow *The physicists*]

Only Fabermacher remained at Erik's shoulder and he was smiling. "They're real!" he murmured. "My God, they are absolutely real!" Erik turned. "Didn't you believe that the neutron existed?" "Oh, I *believed*." Fabermacher shrugged away the phase. "To me neutrons were symbols, n with a mass $m_n = 1.008$. But until now I never *saw* them."
[Mitchell Wilson *Live with lightning*]

If it is not the duty, it should at least be the delight, of poets to contemplate the world of science … Particles we 'know' only by their tracks in the cloud-chamber, light conceived sometimes as particles and sometimes as waves, causality lost in subatomic behaviour, particles 'unobservable' because mere observation imposes changes on the object – how do we relate all this to the solid world of which electrons, atoms and molecules are a part?
[Edwin Morgan   *Essays*]

Neutrinos they are very small.
They have no charge and have no mass
And do not interact at all.
The earth is just a silly ball
To them, through which they simply pass,
Like dustmaids down a drafty hall
Or photons through a sheet of glass.
[John Updike *Cosmic gall*]

Three quarks for Muster Mark!
Sure he hasn't got much of a bark.
And sure any he has it's all beside the mark.
[James Joyce *Finnegans wake*]

To watch so small a thing in motion
As what we've christened the "Higgs boson,"
A tiny, massive thing that passes
For what can best explain the masses
Of other things we cannot see
But somehow, nonetheless, must be.
[Jay Curlin *The New Yorker (*2012)]

The idea that matter is made up of tiny particles - atoms - dates back to antiquity. Atoms were long thought of in the terms laid down by Isaac Newton - 'solid, massy, hard, impenetrable, moveable particles'. This view came to be increasingly questioned during the nineteenth century as the spectra produced by atoms was investigated. It was found that each element produced light at a number of different wavelengths (the particular wavelengths concerned varying with the element). This suggested that atoms were more complex than Newton's description would imply. If so, were atoms the ultimate particles, or did sub-atomic particles exist? The question was answered at the end of the nineteenth century, when J.J. Thomson discovered that atoms contained smaller particles - the electrons. (The name comes from same Greek word as 'electric', and a stream of electrons do, indeed, constitute an electric current.) Thomson showed that his particles had a negative electrical charge. The question was - how much charge did each electron have?

This question was taken up by Robert Millikan in the United States. He devised an apparatus which allowed a charged oil drop to be held suspended by an electric force. This allowed him to measure the electric charge on the drop. He showed that this charge was always some multiple of a specific number which he identified as the electric charge of individual electrons. Millikan's experiments were carried out before the First World War. They formed the basis for student laboratory exercises in subsequent decades. Robert Frost took part in one such exercise [first extract]. Frost's comments are very much to the point. Observations of the oil drops did not always lead to sensible results: so it was often a question of trying to decide which results should be accepted. Atoms are electrically neutral. If electrons have negative charges, atoms must also have some component that is positively charged. Rutherford demonstrated that atoms consist of a cloud of electrons surrounding a small positively charged nucleus. The latter contains positively charged particles - protons - each with a positive charge equal in amount to the electron's negative charge. C.P. Snow, who was acquainted with Rutherford, notes [second extract] that Rutherford's model of the atom meant that one element differed from another due to the different number of protons in its nucleus.

The idea that protons formed the sole constituents of the nucleus soon ran into problems. In particular, nuclei were more massive than would be expected judging from their number of protons. It was soon realised that nuclei contained a second constituent - the neutron - which helped make the nuclei stable. As the name implies, neutrons have no electrical charge. Mitchell Watson - who was, for a time, a research physicist - notes that a neutron is very slightly heavier than a proton[third extract]. Particles with an electric charge are relatively easy to detect: neutral particles much less so. The universal method of detecting particles in the early days was the cloud chamber [fourth extract]. A charged particle passing through this left a trail of water droplets. Neutral particles did not, and could initially only be detected indirectly. Hence the joy in the third extract at a new way of detecting them.

This simple picture of sub-atomic particles soon became more complicated as more and more particles were discovered - the neutrino, for example. The existence of this particle was asserted on the basis of theory, rather than experiment. It was supposed to have no mass and no charge and, consequently, very little interaction with anything else. As Updike [fifth extract] says, most neutrinos hitting the Earth pass straight through it without interaction. (Actually things have become more complicated in recent years, as the neutrino has been found to have a very small mass and to come in more than one form.)

Another complication has been the realisation that particles such as protons and neutrons are actually made up of still smaller particles. These are called *quarks*. The reason can be found in the sixth extract from Joyce. This is an unusual example of the scientist taking something from the writer, rather than vice versa. It was initially believed that there were three quarks, so the first line seemed appropriate. There are now more: so the name is less appropriate, but has been retained. (A 'quark' is actually a cawing noise, like that of a crow.) The Higg's boson [final extract] brings us right up to date. It was predicted many years ago as a way of understanding how sub-atomic particles have acquired their masses. Its likely experimental detection was announced in March 2013.

Mr. Peter above them both, who after dinner did show us the experiment (which I had heard talk of) of the chymicall glasses, which break all to dust by breaking off a little small end; which is a great mystery to me. [Pepys' *Diary* 13 January 1662]

There comes also Mr. Reeve, with a microscope and scotoscope. For the first I did give him £5 10s., a great price, but a most curious bauble it is, and he says, as good, nay, the best he knows in England, and he makes the best in the world. The other he gives me, and is of value; and a curious curiosity it is to look objects in a darke room with. Mightly pleased with this I to the office, where all the morning. ...... then to read a little in Dr. Power's book of discovery by the Microscope to enable me a little how to use and what to expect from my glasse. [Pepys' *Diary* 13 August 1664]

But by and by comes Mr. Reeves, and after him Mr. Spong, and all day with them, both before and after dinner, till ten o'clock at night, upon opticke enquiries, he bringing me a frame he closes on, to see how the rays of light do cut one another, and in a darke room with smoake, which is very pretty. He did also bring a lanthorne with pictures in glasse, to make strange things appear on a wall, very pretty. We did also at night see Jupiter and his girdle and satellites, very fine, with my twelve-foote glasse, but could not Saturne, he being very dark. Spong and I had also several fine discourses upon the globes this afternoon, particularly why the fixed stars do not rise and set at the same houre all the yeare long, which he could not demonstrate, nor I neither, the reason of. [Pepys' *Diary* 19 August 1666]

I took coach, having first discoursed with Mr. Hooke a little, whom we met in the streete, about the nature of sounds, and he did make me understand the nature of musicall sounds made by strings, mighty prettily; and told me that having come to a certain number of vibrations proper to make any tone, he is able to tell how many strokes a fly makes with her wings (those flies that hum in their flying) by the note that it answers to in musique during their flying.[Pepys' *Diary* 8 August 1666]

Dr. Wilkins saying that he hath read for him in his church, that is poor and a debauched man, that the College have hired for 20s. to have some of the blood of a sheep let into his body; and it is to be done on Saturday next. They purpose to let in about twelve ounces; which, they compute, is what will be let in in a minute's time by a watch. They differ in the opinion they have of the effects of it; some think it may have a good effect upon him as a frantic man by cooling his blood, others that it will not have any effect at all. [Pepys' *Diary* 21 November 1667]

The King's medicinal Garden and Laboratory with all the Apparatus, is at no hand to be omitted, because it is so well furnish'd and so rarely fitted for the Design, as having all the affections of Ground and Situation desirable. If you stayd a whole Winter in Paris, I would invite you to see a Course of Chimistry, which is both there and in several private Places shew'd to the Curious to their wonderful satisfaction and Benefit of Philosophic Spirits. And if his Majesty have done any thing for the Virtuosi (our Emulators) in designing them a Mathematical College, seek after it, and procure to be admitted into their present Assembly, that you may render our Society an Account of their Proceeding. You will easily obtain that by the assistance of some Friend: But Mr Oldenburg being in the Country, (for I went to his House) you will miss of an infallible Address.[Letter from John Evelyn to Pepys (21 August 1669) advising him what to see in Paris]

I take it to be the same case whether a man, to throw sixes have one throw with twelve dyes or two throws with six, but I reccon it an easier task to throw with six dyes one six at one throw than two sixes at two throws.[Letter from Isaac Newton to Pepys 23 December 1693]

Pepys   DISCUSSION

Samuel Pepys' diary covers the 1660s, an especially significant time for the development of science in England. The Royal Society received its charter in 1662 and rapidly became the main focus for scientific discussion in the capital. Pepys, curious about everything, soon became a member, and his diary often reflects the matters that then concerned the Society. Thus the first extract reflects the contemporary interest in Prince Rupert's drops [the 'chymicall glasses']. Prince Rupert was a cousin of Charles II and brought some samples of these 'drops' from the Continent in 1660. They were made by dropping liquid glass into water, so producing a bead of glass with a long tail. Breaking the latter led to loud bang. The question, which has only been fully answered recently, was why? The telescope and the microscope both appeared around the beginning of the seventeenth century. By the mid-century, instrument-makers in London were producing samples of both. Richard Reeves [second extract] was a leading maker, so it was natural for Pepys to turn to him. (There is some doubt what the 'scotoscope' was, but it may have provided illumination for the microscope when there was little light.) Pepys was very up-to-date in his reading: Henry Power's book on *Experimental philosophy* had appeared earlier the same year. Power, an early member of the Royal Society, based his observations on instruments made by Reeves.

John Spong, another practitioner, was interested in optical phenomena, as is evident from the third extract. Rays of light are normally invisible, but if dust particles are present, the light scattered by the particles allows the path of a ray to be followed. In a darkened room, the path of a ray can then be traced as it is reflected from surfaces. Magic lanterns were a seventeenth-century invention. As the name suggests, they could be used for magical purposes - by producing images of ghosts, for example. The size of telescopes nowadays is usually measured by the diameter of their main lens or mirror. In early days they were measured by their focal length. (The simple lenses available in the Pepys' time produced distorted images; these could be improved, to some extent, by having a long focal length.) The most revealing sentence is the last. Schools taught very little mathematics in Pepys' day: he picked up all he knew on the job. It seems that Spong was little better at solving this relatively easy problem.

Robert Hooke [fourth extract] was one of the great scientific figures of the 1660s. In the latter part of the 1650s, he had shown his skills while assisting Robert Boyle in his experiments. When the Royal Society was founded, he was offered the post of curator, with the requirement that he provided new scientific experiments or observations for discussion at its meetings. Pepys enjoyed his contacts with Hooke: he was particularly impressed by Hooke's *Micrographia*. Early in 1665, he noted that he had bought a copy and sat up late reading 'the most ingenious book that I ever read in my life'. The group round Dr. Wilkins at Oxford in the latter part of the 1650s had developed an interest in the idea of blood transfusion. Indeed, one of them, Christopher Wren, produced an early attempt at a hypodermic syringe. After the Restoration, they transferred this activity to the Royal Society in London, where it aroused Pepys' interest. For the most part, these experiments were done with animals, but Pepys [fifth extract] reports on the first experiment with a human being. (The man involved seems to have been a somewhat eccentric Cambridge graduate.) The subject claimed to have felt better afterwards, though Pepys noted that his character seemed little changed from what it had been.

The other great diarist of the 1660s, John Evelyn, was in contact with Pepys for two reasons. The first was their involvement in the care of sick seamen; the second was their involvement with the Royal Society. In the sixth extract, Evelyn is advising Pepys what was worth seeing in the scientific world of Paris. The Academy of Sciences was set up in Paris in 1666 under the patronage of Louis XIV. The slightly later date explains why Evelyn refers to them as 'our [the Royal Society's] emulators'. The Paris Academy was from the start a state institution (unlike the Royal Society), and so was involved in Royal enterprises. Henry Oldenburg was a German, living in London, who became the first Secretary of the Royal Society. He was particularly involved in exchanging information with the Paris Academy, so would have been a valuable contact for Pepys, if available. All the extracts so far have been from the 1660s, when Pepys was keeping his diary. The final extract comes from a good deal later. Pepys served as President of the Royal Society from 1684 to 1686. At the time, books had to be given clearance before they could be published; the President of the Royal Society was one of the people with the power to give such clearance. Consequently, when Newton's *Principia* was published in 1687, Pepys' authorisation appears at the bottom of the title page [see the image at the beginning of *Quotations*]. This was not Pepys only contact with Newton. As the final extract indicates, they exchanged letters about a question of probability that arose when gambling with dice. This is still known as the 'Newton-Pepys problem' in probability circles.

The harmony of things, – this hard decree,
This uneradicable taint of sin,
This boundless upas, this all-blasting tree,
Whose root is earth
[Lord Byron *Childe Harold's pilgrimage*]

He often informed Mrs Todgers that the sun had set upon him; that the billows had rolled over him; that the car of Juggernaut had crushed him, and also that the deadly Upas tree of Java had blighted him.
[Charles Dickens *Martin Chuzzlewit*]

"What's the tree I read about somewhere that does you in if you sit under it?", to which Jeeves replies, "The upas tree, sir." Bertie rejoins, "She's a female upas tree. It's not safe to come near her."
[P.G. Wodehouse *Stiff Upper Lip, Jeeves*]

..... three died from eating the poisonous herb called "tulip." Five more sickened from this cause, but we managed to cure them with doses of an infusion made by boiling down the tulip leaves. If administered in time this is a very effective antidote.
[H. Rider Haggard *King Solomon's mines*]

They poured strychnine in his cup
And shook to see him drink it up:
They shook, they stared as white's their shirt:
Them it was their poison hurt.
—I tell the tale that I heard told.
Mithridates, he died old.
[A.E Housman *Terence, this is stupid stuff*]

After a time I crawled home, took some food and a strong dose of strychnine, and went to sleep in my clothes on my unmade bed. Strychnine is a grand tonic, Kemp, to take the flabbiness out of a man.
[H.G. Wells *The invisible man*]

Suppose you knew beforehand the poison that would be made use of against you; suppose the poison was, for instance, brucine" --
"Brucine is extracted from the false angostura is it not?" inquired Madame de Villefort.
[Alexandre Dumas *The Count of Monte Cristo*]

If the child were pricked with one of those arrows dipped in curare or some other devilish drug, it would mean death if the venom were not sucked out.
[A. Conan Doyle *The adventure of the Sussex vampire*]

I believe that, save for one sample in a laboratory at Buda, there is no other specimen in Europe. It has not yet found its way either into the pharmacopoeia or into the literature of toxicology. The root is shaped like a foot, half human, half goatlike; hence the fanciful name given by a botanical missionary. It is used as an ordeal poison by the medicine-men in certain districts of West Africa and is kept as a secret among them.
[A. Conan Doyle *The adventure of the devil's foot*]

No hitherto undiscovered poisons may be used, nor any appliance which will need a long scientific explanation at the end.
[R.A. Knox Preface to *Best Detective Stories of 1928-29*]

Early European voyages of exploration brought back a number of plants, some of which were poisonous (maybe we should, in retrospect, add tobacco to that list). However, the systematic retrieval and classification of plants from abroad started around the time of Captain Cook, towards the end of the eighteenth century, when trained naturalists began to accompany the expeditions. The Victorians, with their combined concern for plants, exploration and death, were particularly interested in poisonous plants, and often refer to them.

One plant regarded as especially virulent was the upas tree. Though known earlier, it came to European attention in the latter part of the eighteenth century, when accounts of its alleged deadly influence on its surroundings were reported. The story was quickly picked up by Erasmus Darwin, who reported in *The loves of plants* that: 'there is a poison-tree in the island of Java, which is said by its effluvia to have depopulated the country for twelve or fourteen miles'. The upas tree's supposed effects made an impact on the popular imagination which lasted - as the first three extracts indicate - through the nineteenth century and into the twentieth. Nor was this restricted to the English-speaking world: there is a Pushkin poem which says much the same as Erasmus Darwin. Although the original report specified the tree as growing in Java, it actually spreads across Asia and Africa. Though hardly as ferocious as its reputation, the sap of the tree has long provided a poison for arrows.

If the upas tree was linked with Java, the cape tulip, as its full name suggests, was linked with southern Africa. The plant contains chemicals that affect the heart, so the statistics contained in the fourth abstract are hardly surprising. The tulips have a similar effect on other animals, including cattle. (Unfortunately, the plant's attractive appearance led to its importation into Australia to the disadvantage of the farmers there.) The interesting point in this abstract is the use of a dilute solution of the plant to offset its original effects. The idea that small doses of a poison can protect against large doses is both old and widespread. The most famous proponent of this approach was the Mithridates [actually Mithridates VI - it was a common name] celebrated by Housman in the fifth abstract. He is remembered as one of the most successful (and long-lived) opponents of the Roman Republic.

Strychnine comes from the seeds of another tree native to Indonesia. It has been a popular plant poison since antiquity - as witness Mithridates - but throughout the nineteenth century it was promoted, when taken in small doses, as a health tonic. It was thought both to stimulate the digestion and to improve muscle activity. The latter property was presumably why it helped Wells' invisible man fight against flabbiness. Brucine is a chemical closely related to strychnine, but less poisonous. It was used for heart problems, especially high blood pressure. Its mention in the Dumas' novel is related to the Mithridates legend.

Poison arrows, whether used in Asia, Africa, or South America, fascinated the Victorians; The name *curare* actually covers a group of poisons found in South America. They have the unusual property that they only act when introduced into the blood stream - as with an arrow. They can be eaten without producing any harm. This is why the mother in the eighth abstract can suck out the poison without being harmed herself. Sherlock Holmes is clearly not sure whether curare is the poison that has been used here. Subsequent commentators have suggested that other South American poisons might fit the symptoms he describes better.

During his days as a medical student, Conan Doyle took a special interest in plant preparations (as an examination of his heavily annotated *materia medica* textbook has shown). He later visited West Africa [ninth abstract] and learnt about the arrow poisons used there. It is not certain which plant the 'devil's foot' was supposed to be, but it might be Strophanthus. The seeds of this plant provided an oil used as an arrow poison in West Africa - it affects the heart. The plethora of new mysterious poisons found in the tropics was a godsend for writers of thrillers and detective stories. At the end of the 1920s, R.A. Knox introduced ten rules intended to provide fair play for readers of detective stories. The final abstract contains the fourth rule of these rules. Knox was a Roman Catholic priest and writer of detective stories. (His novels typically include the implicit additional rule that no Roman Catholic suspect could actually be the criminal.)

Vast chain of being! which from God began,
Natures ethereal, human, angel, man,
Beast, bird, fish, insect, what no eye can see,
No glass can reach; from Infinite to thee,
From thee to nothing. On superior powers
Were we to press, inferior might on ours:
Or in the full creation leave a void,
Where, one step broken, the great scale's destroyed:
From Nature's chain whatever link you strike,
Tenth or ten thousandth, breaks the chain alike.
[Alexander Pope *An essay on man*]

... this scale of being I have demonstrated to be raised by presumptuous imagination, to rest on nothing at the bottom, to lean on nothing at the top, and to have vacuities from step to step through which any order of being may sink into nihility without any inconvenience, so far as we can judge, to the next rank above or below it.
[Samuel Johnson A review of *A free inquiry into the nature and origin of evil*]

        .... it will one day be found
With other relics of 'a former world',
When this world shall be former, underground,
Thrown topsy-turvy, twisted, crisp'd, and curl'd ....
So Cuvier says:- and then shall come again
Unto the new creation, rising out
From our old crash
[Lord Byron *Don Juan Canto IX*]

From the mute shell-fish gasping on the shore,
To men, to angels, to celestial minds,
Forever leads the generations on
To higher scenes of being
[Mark Akenside *Pleasures of the imagination*]

A subtle chain of countless rings
The next unto the farthest brings;
The eye reads omens where it goes,
And speaks all languages the rose;
And, striving to be man, the worm
Mounts through all the spires of form.
[Ralph Waldo Emerson *Nature*]

Doubtless many will reply that they can more easily conceive ten millions of special creations to have taken place, than they can conceive that ten millions of varieties have arisen by successive modifications. All such, however, will find, on inquiry, that they are under an illusion...... If they have formed a definite conception of the process, let them tell us how a new species is constructed, and how it makes its appearance. Is it thrown down from the clouds? or must we hold to the notion that it struggles up out of the ground? Do its limbs and viscera rush together from all points of the compass?
[Herbert Spencer *The development hypothesis*]

The idea of a 'chain of being' dates back to classical times. It supposes that we live in an orderly universe in which everything has its proper place. As Pope says in the first extract, the chain goes all the way from God at the top to nothing at the bottom. En route, it passes through angelic beings, humans, animals, plants and minerals. Each link in the chain is seen as being superior to what is below it and inferior to what is above it. The classification scheme - for that is what it is - goes into much finer detail than this. Thus birds were put below mammals, but above fishes; in turn, birds were divided into groups, with birds of prey at the top and seed-eating birds at the bottom. Pope stresses two points about the chain. Firstly, the chain is continuous, with no gaps between groups. Secondly, any attempt to change it would be disastrous. (Compare the Shakespearian comment: 'Take but degree away, untune that string, And, hark, what discord follows!') This static chain of being was widely accepted in the eighteenth century, but it did not go unchallenged. There was, for example, the criticism that Samuel Johnson propounded [second extract]. How could a chain be continuous if each link showed clear differences from the next? Surely, this left spaces for any number of hybrids in between. This also meant, he reckoned, that it should be possible to remove a link from the chain without the whole chain collapsing.

By the early nineteenth century, studies of fossils had shown that forms of life had existed on Earth in the past which are no longer present today. The question was how this squared with the chain-of-being notion. Georges Cuvier, who was then the leading figure in research on fossils believed that the Earth had undergone a series of catastrophes. At each catastrophe life had been wiped out, only for new forms to be created in its aftermath. Between each catastrophe the life forms present remained constant. Byron, an interested, if amused, spectator of science, lays out the idea succinctly [third extract]. Cuvier agreed with Johnson that the idea of a chain of being was unsupportable. However, his work does allow the possibility for a new chain to come into existence after each extinction.

Looked at in another way, the chain of being could lead to thoughts of evolution. Humans certainly wished to progress up the chain. Might this not be true of all the other links in the chain? In other words, might not the chain be dynamic, rather than static? This possibility was already being discussed [fourth extract] at the time that Pope was writing his *Essay on man*. (Pope actually read and commended Akenside's poem prior to its publication.) In the nineteenth century, such a belief in progress became increasingly acceptable. Emerson published his essay on nature, to which the poem [fifth extract] is attached, in the 1830s. It launched the transcendentalist movement, which pressed for a quasi-religious belief in nature and in the possibility of improvement.

In the following decade, Robert Chambers, a Scottish publisher, took the concept further. His book *Vestiges of the natural history of creation* proclaimed that the unifying theme in nature was development. So far as life on Earth was concerned, this meant a progressive transmutation of species. The book became very popular - Prince Albert read it aloud to Queen Victoria - but it also soon came under attack. It was naturally suspect to the clergy, but scientists, too, were unhappy with it. Charles Darwin commented: '... the writing & arrangement are certainly admirable, but his geology strikes me as bad, & his zoology far worse'. Chambers was associated with a salon, run by his fellow-publisher, John Chapman, which included a number of contemporary radical thinkers. Another member was Herbert Spencer, one of the founding fathers of sociology. Spencer developed a close friendship with George Eliot who, along with G.H. Lewes, also belonged to the group. All three accepted the idea of progressive development. Spencer believed not only that nature was governed by evolution, but that such evolution was the result of a universal natural law. The final extract - from an essay published just four years before Darwin's *Origin of species* - shows him ridiculing earlier ideas. In the 1860s, after he had read Darwin's book, Spencer coined the phrase 'survival of the fittest', which Darwin, himself, was using by the end of the decade. Unlike Darwin, however, Spencer believed evolution was a directional process, with lower forms of life inevitably leading on to higher forms.

Pure mathematics was discovered by Boole in a work which he called the *Laws of Thought* (1854). This work abounds in asseverations that it is not mathematical, the fact being that Boole was too modest to suppose his book the first ever written on mathematics. He was also mistaken in supposing that he was dealing with the laws of thought ….. His book was in fact concerned with formal logic, and this is the same thing as mathematics. [Bertrand Russell *Mathematics and the metaphysicians*]

When he was young his cousins used to say of Mr Knight
'This boy will write an algebra – or looks as if he might.'
And sure enough, when Mr Knight had grown to be a man,
He purchased pen and paper and an inkpot, and began.[E.V. Rieu *Hall and Knight*]

I tell my students the story of Newton versus Leibniz,
the war of symbols, lasting five generations,
between The Continent and British Isles,
involving deeply hurt sensibilities,
and grievous blows to national pride;
on such weighty issues as publication priority
and working systems of logical notation [Sarah Glaz *Calculus*]

I thought you were finding a proof for Fermat's last theorem.
It is very difficult. You will have to show me how.
If I knew how, there would be no need to ask *you*. Fermat's last theorem has kept people busy for a hundred and fifty years .[Tom Stoppard *Arcadia*]

Time is *a* fourth dimension, but I'm thinking about a fourth spatial dimension, like length, breadth and thickness. For economy of materials and convenience of arrangement you couldn't beat it. To say nothing of the saving of ground space - you could put an eight-room house on the land now occupied by a one-room house. Like a tesseract. [Robert A. Heinlein *And he built a crooked house*]

Pathological monsters! cried the terrified mathematician
Every one of them is a splinter in my eye
I hate the Peano Space and the Koch Curve I fear the Cantor Ternary Set
And the Sierpinski Gasket makes me want to cry
And a million miles away a butterfly flapped its wings
On a cold November day a man named Benoît Mandelbrot was born
[Jonathan Coulton *Mandelbrot set*]

Nothing on the map to the contrary. Only a classical Poisson distribution, quietly neatly sifting among the squares exactly as it should ... growing to its predicted shape ...
"But squares that have already *had* several hits, I mean -"
"I'm sorry. That's the Monte Carlo Fallacy. No matter how many have fallen inside a particular square, the odds remain the same as they always were. Each hit is independent of all the others."
[Thomas Pynchon *Gravity's rainbow*]

Pure mathematics   **DISCUSSION**

The exact nature of basic mathematical concepts (such as number, function, and so on) has been the subject of discussion for many centuries. However, arguments about the foundations of mathematics became especially lively at the beginning of the twentieth century. Bertrand Russell was a leading proponent at that time of the view that mathematics is simply a form of logic. In the first extract, he notes that a British mathematician, George Boole, pioneered the systematic study of mathematical logic, and so - he asserts - of the real foundations of mathematics. Boole, as Russell explains, made less grandiose claims. In the event,

Boole's work has proved basic to the development of computing, but Russell's wider claims have received less backing. Pure mathematics textbooks often go through many editions and have a long life. Hall and Knight's *Elementary algebra for schools* is a good example. It was first published in the nineteenth century and continued in use until well after the Second World War. Its sequel, *Higher algebra*, is still a recommended text in India. Since the titles of mathematics textbooks tend to be rather similar, the books are usually referred to by the names of their authors. Rieu will have used the elementary text at school [second extract]. He is best remembered as the founder editor and translator of the Penguin Classics series, but he also wrote verse for children.

Algebra, which represents unknown quantities by symbols, is one of the dominant fields in mathematics. Sarah Glaz is an algebra specialist: an American mathematics professor who also writes poetry. The third extract deals with the controversy that accompanied the birth of calculus. In the 1660s, Isaac Newton began to devise a form of calculus that he subsequently used for studying the motion of bodies. He was always slow to publish his work, and did not provide full details until forty years later. Meanwhile, Gottfried Leibniz in Germany began working on a different approach to calculus in the 1670s. Unlike Newton, he published his results. This led to a major debate - to some extent still continuing today - on which of them should be accorded priority. In terms of the notation used, that devised by Leibniz has proved to be the winner.

Fermat was a seventeenth-century French mathematician who contributed to a number of fields. His last theorem concerned whole numbers. Pythagoras' theorem says that there are certain whole numbers for which $a^2 + b^2 = c^2$ Fermat asserted that this equation could not be true for any index higher than two. (He claimed to have a proof, but failed to provide it. So his assertion was actually a conjecture, rather than a theorem.) Many attempts were made down the years to provide a proof. All failed until, in 1995, the English mathematician, Andrew Wiles, finally provided one. Tom Stoppard's play *Arcadia* [fourth extract] was first performed just two years before this proof.

The latter part of the nineteenth century saw a growing interest in multi-dimensional geometry. The space in which we live has three dimensions. What would geometry be like in a higher-dimensional space? The word 'tesseract' was coined at that time to describe the four-dimensional equivalent of the cube. The possibilities offered by such complex space naturally attracted the attention of science-fiction authors. Robert Heinlein was one of the leading American science-fiction authors in the golden years after the Second World War. The fifth extract comes from an amusing short story by him about architecture in four-dimensional space. The sixth extract is replete with the names of mathematicians - Peano, Koch, Cantor and Sierpinski - of the late nineteenth or early twentieth century, who worked on particular kinds of curves. These curves can be related to the idea of a *fractal* - a curve that has similar properties at whatever scale you examine it. Mandelbrot coined the word in the 1970s, and applied the idea in a variety of contexts. Fractals relate to another recent area of interest in mathematics - chaos theory. This studies systems that are very sensitive to the initial conditions. The 'butterfly effect' is a memorable image of such sensitivity. It suggests that the formation of a hurricane may depend on whether or not a butterfly has flapped its wings at an earlier time. Coulton is a singer and songwriter - a former undergraduate at Yale, where Mandelbrot was professor. Mandelbrot apparently enjoyed this song.

Statistics is nowadays considered a field of study in its own right, but it started life as a branch of mathematics. The nineteenth-century French mathematician, Poisson, provided a theoretical basis for estimating the probability that a particular number of events would occur over a specific interval of time or space. Pynchon introduces the concept via a discussion of German V-2 missile impacts in the Second World War. The Monte Carlo fallacy is also known as the gambler's fallacy. It is the erroneous belief that, if an event currently occurs more often than expected, then it will come up less often in the future. Thus, if a particular number comes up unusually often at a game of roulette, the gambler assumes that it will occur less often in the rest of the session.

And the rockets' red glare, the bombs bursting in air,
Gave proof through the night that our flag was still there;
[Francis Scott Key *The Star-Spangled Banner*]

What is glory? - in the socket
See how dying tapers fare!
What is pride? - a whizzing rocket
That would emulate a star.
[William Wordsworth *Inscriptions supposed to be found in and near a hermit's cell*]

"Once the rockets are up, who cares where they come down?
That's not my department," says Wernher von Braun.
[Tom Lehrer *Wernher von Braun*]

Till from the south-west, as their low scream mars
And halts this warm hypnosis of the dark,
Three black automata cut swift and stark,
Shaped clearly by the backward flow of stars.
Stronger than lives, by empty purpose blinded,
The only thought their circuits can endure is
The target-hunting rigour of their flight;
[Robert Conquest *Guided missiles experimental range*]

How many more years I shall be able to work on the problem I do not know; I hope, as long as I live. There can be no thought of finishing, for 'aiming at the stars' both literally and figuratively, is a problem to occupy generations, so that no matter how much progress one makes, there is always the thrill of just beginning.
[Robert H. Goddard *Letter to H. G. Wells in 1932*]

Indeed the early history of rocket design could be read as the simple desire to get the rocket to function long enough to give an opportunity to discover where the failure occurred. Most early debacles were so benighted that rocket engineers could have been forgiven for daubing the blood of a virgin goat on the orifice of the firing chamber.
[Norman Mailer   *Of a fire on the Moon*]

I snuggled down into the couch - and then, without any warning, felt as if someone had jumped on top of me. There was a tremendous roaring in my ears, and I seemed to weigh a ton. It required a definite effort to breathe - that was no longer something you could leave to your lungs and forget all about. The feeling of discomfort lasted only a few seconds: then I grew accustomed to it. The ship's own motors had not yet started: we were climbing under the thrust of the booster rockets, which would burn out and drop away after thirty seconds.
[Arthur C. Clarke *Islands in the sky*]

Science, she would like to know,
In her complacent ministry of fear,
How we propose to get away from here
          ..... Will she be asked to show
Us how by rocket we may hope to steer
To some star off there, say, a half light-year
[Robert Frost *Why wait for science?*]

By the thirteenth century, Western Europeans knew both of the Chinese invention of gunpowder, and of its use in rockets. In subsequent centuries, rockets became recognised, if not widely used, weapons of war. Towards the end of the eighteenth century, more powerful rockets were developed in Mysore in India and used against the forces of the East India Company there. Samples brought back to Britain formed the basis for a new military rocket developed by William Congreve. Earlier rockets had been highly unstable in flight; Congreve attached a long stick that led to improved stability. (This is the method still used with fireworks.) Congreve rockets were used at the battle of Baltimore in 1814, leading to the well-known line in the US national anthem [first extract]. Though civilians in Britain were aware of this military use, their own acquaintance with rockets usually came from the public fireworks displays that were popular from the latter part of the eighteenth century onward. It is likely that it was this that Wordsworth had in mind when he wrote the second extract.

Solid fuel could be used for relatively small rockets, but was difficult to control and use for larger rockets. So, in the twentieth century, much effort was put into developing liquid-fuel rockets. This came to a peak in the Second World War, when the Germans under Wernher von Braun developed the V2 rocket which had a range of some 300Km and a reasonable guidance system. At the end of the war, von Braun and a number of his colleagues were captured by US forces and taken to the USA to help develop rocketry there. This American incarnation of a former Nazi party member inspired the song by the mathematician and satirist, Tom Lehrer.

After the war, the USA was not the only country to develop guided missiles. The main UK base for testing such missiles was set up in the outer Hebrides during the 1950s, when Conquest's poem was written [fourth extract]. Robert Conquest is a well-regarded poet, but is even better known as a historian of the Soviet Union, who explored the full scale of the repression there. By the end of the 1950s, it became evident - as the space race started - that the USSR was ahead of the USA in the development of guided missiles.

In a sense, guided missiles were a side-effect of the desire to use rockets for space travel. A film about von Braun had him saying: 'I aim at the stars, but sometimes I hit London'. Between the wars, one of the leading pioneers of liquid-fuel rockets was Robert Goddard in the USA. With the support particularly of the Smithsonian Institution, he launched some thirty rockets during the 1930s. His interest in space exploration had been roused by reading H.G. Wells' *The war of the worlds* as a boy, and as the fifth extract indicates, he corresponded with Wells when he later began to launch his own rockets.

The 1950s saw the appearance of 'New Journalism' - writing factual stories in the style of literary fiction. Norman Mailer was one of its leading proponents. Mailer, who had studied aeronautical engineering at Harvard, became fascinated by the American space programme. *Of a fire on the Moon* first appeared in the magazine *Life*, then as a book in 1970. It examines the events leading up to the Apollo 11 landing on the Moon. Mailer makes clear that the initial stages of the programme were plagued by rocket failures [Extract 6]. Clarke, who was a radar specialist during the Second World War, started writing about space travel soon after the war. He, Asimov and Heinlein were often referred to as the 'Big Three' of science fiction in later years (with Clarke best known for the film *2001*). *Islands in the sky* was one of Clarke's earliest stories: it appeared in 1952 - before the beginning of the space age. In consequence, he felt it necessary to explore the process of getting into space in more detail than most later S-F stories. Here, Clarke is imagining what happens at launch.

Robert Frost had a love/hate relationship with science: it fascinated him, but he also feared it. This comes over in the final extract. If science destroys the Earth, we will yet have to use science - in this case rocket power - to escape. What Frost does not say is that it would take many human generations for a rocket-powered spacecraft to reach the nearest star (and the problem is worse than he says, since the nearest star is over four light-years away, not half a light-year.)

If it form the one landscape that we, the inconstant ones,
Are consistently homesick for, this is chiefly
Because it dissolves in water. Mark these rounded slopes
With their surface fragrance of thyme and, beneath,
A secret system of caves and conduits; hear the springs
That spurt out everywhere with a chuckle,
Each filling a private pool for its fish and carving
Its own little ravine
[W.H. Auden *In praise of limestone*]

The water abrades,
Erodes; dissolves
Limestones and chlorides
[Norman Nicholson *Beck*]

Above the collar of crags,
The granite pate breaks bare to the sky
Through a tonsure of bracken and bilberry.
[Norman Nicholson *Eskdale Granite*]

The composition of the huge hill was revealed to its backbone and marrow here at its rent extremity. It consisted of a vast stratification of blackish-gray slate, unvaried in its whole height by a single change of shade.
[Hardy *A pair of blue eyes*]

Then Abner Dean of Angel's raised a point of order, when
A chunk of old red sandstone took him in the abdomen,
And he smiled a kind of sickly smile, and curled up on the floor,
And the subsequent proceedings interested him no more.
[Bret Harte *Plain language from Truthful James*]

And first, therefore, let us ask what questions imperatively need answer, concerning indisputable lavas, seen by living human eyes to flow incandescent out of the earth, and thereon to cool into ghastly slags. On these I have practically burnt the soles of my boots, and in their hollows have practically roasted eggs; and in the lee of them, have been wellnigh choked with their stench ; and can positively testify respecting them, that they were in many parts once fluid under power of fire, in a very fine and soft flux ; and did congeal out of that state into
ropy or cellular masses, variously tormented and kneaded by explosive gas.
[Ruskin *Deucalion*]

Weep not, good reader! He is truly blest,
Amidst chalcedony and quartz to rest—
Weep not for him! but envied be his doom,
Whose tomb, though small, for all he loved had room
And, O ye rocks! schist, gneiss, whate'er ye be,
Ye varied strata, names too hard for me,
Sing 'O be joyful!' for your direst foe,
By death's fell hammer, is at length laid low.
[Felicia Dorothea Hemans *Epitaph on a mineralogist*]

Mentions of rocks abound in literature, but some writers seem to have an especial interest in them. Auden is one. He wrote *In praise of limestone* - often seen as one of his best poems - while living in Italy after the Second World War. The poem itself has been interpreted as an allegory - for Mediterranean civilisation, or for the human body - but the poem can also be read more literally. Limestone is a rock found not only in the Mediterranean, but also in Yorkshire, where Auden was born. Indeed, in a letter he sent while writing this poem, Auden noted that he had not previously realised how similar the Mediterranean landscape was to the Pennines. It is hardly surprising that writers find limestone particularly interesting. The solubility of limestone in water leads to the creation of complex channels and cavities (often including extensive cave systems) in the rock, producing what are called 'karst' landscapes. (The word, *karst*, derives from the name of a limestone area north of Trieste.)

Another mention of this water erosion of limestone occurs in the second extract. Nicholson lived all his life in Cumbria, in an area where there was much mining and quarrying, so it was natural for him to be interested in the local rocks. (He was also influenced in terms of poetry by his contemporary, Auden.) The first brief extract notes that limestone is not the only rock to erode easily. A chloride such as sodium chloride [common salt] erodes even more readily. Along with limestone, Nicholson's surroundings included many areas of granitic rock. Limestone is a sedimentary rock: it is laid down from water at ordinary temperatures, and often contains the shells of microfossils. Granite, by contrast, is an igneous rock, which formed at a high temperature, and is very hard. In consequence, granite erodes slowly, and, as in the third extract, tends to stand out from its surroundings.

Hardy had a fondness for hills in his novels. The next extract, from one of his early stories, is set around the village of Boscastle on the north coast of Cornwall. One of the major exports from Boscastle for centuries had been slate; so Hardy, in this novel, naturally invokes a slate hill. Slate is an example of a metamorphic rock: one that has been altered from its original state by heat and pressure. It ranks as a category alongside the sedimentary and igneous rocks of the previous extracts. In origin, much slate comes from mud, deposited perhaps by rivers, which is later compressed and heated. Sandstone has a rather similar origin, though it consists of larger particles. These are welded together by pressure, but not heated; so sandstone is a sedimentary rock. The American West has massive deposits of sandstone. Unsurprisingly, they figure in many accounts of the pioneering days. Bret Harte was born on the East Coast of the USA, but moved as a teenager to California in 1853. There he became nationally famous for his descriptions of pioneering life. He was also a noted humorist. The extract quoted here reflects both of these characteristics. It is part of a description of a meeting of an imagined scientific society which breaks up into a fight. The weapon of choice is the local rock - old red sandstone - the most famous of all the sandstone deposits.

Rocks typically take some time to form, so what we usually see is the end-product. Flows of lava from volcanoes are an exception to this general rule. Ruskin, in the sixth extract, describes his own experience of seeing them form. Though Ruskin is generally classified as an art critic, he was from his early years fascinated by geology. Later in life, he had one of the best collections of agates - a mineral often associated with volcanic activity - in Britain. Ruskin believed that a knowledge of geology was essential for a landscape painter: a major section of his influential book, *Modern painters*, was actually devoted to geology.

Felicia Dorothea Hemans is no longer a well-known name. Yet her poetry was admired by her contemporary, Wordsworth, who wrote a memorial verse on her death. She is still remembered for her poem *Casabianca* with its initial lines: 'The boy stood on the burning deck/ Whence all but he had fled'. The final extract demonstrates her knowledge of rocks. Thus chalcedony is paired with quartz of which it is a form, while schist and gneiss are both metamorphic rocks. When Hemans wrote this poem in the early years of the nineteenth century, interest in the geologic record, and especially in strata, was rising rapidly: the Geological Society of London was founded in 1807, while the British Geological Survey appeared on the scene in 1835, the year of her death.

A Clerk ther was of Oxenford also,
That unto logyk hadde longe ygo…..
For hym was levere have at his beddes heed
Twenty bookes, clad in blak or reed,
Of Aristotle and his philosophie, [Geoffrey Chaucer *Canterbury tales*]

The fair philosopher to Rowley flies,
Where, in a box, the whole creation lies …..
Of Desaguliers she bespeaks fresh air;
And Whiston has engagements with the fair. [Edward Young *Satire V: On women*]

I purpose … .. to bring before you, in the course of these lectures, the Chemical History of a Candle…..
There is not a law under which any part of this universe is governed which does not come into play, and is touched upon in these phenomena. There is no better, there is no more open door by which you can enter into the study of natural philosophy, than by considering the physical phenomena of a candle.
[Michael Faraday *The chemical history of a candle*]

'Bitzer,' said Thomas Gradgrind. 'Your definition of a horse.'
'Quadruped. Graminivorous. Forty teeth, namely twenty-four grinders, four eye-teeth, and twelve incisive. Sheds coat in the spring; in marshy countries, sheds hoofs, too. Hoofs hard, but requiring to be shod with iron. Age known by marks in mouth.' Thus (and much more) Bitzer.
'Now girl number twenty,' said Mr. Gradgrind. 'You know what a horse is.'
[Charles Dickens *Hard times*]

The tyranny of the past, many think, weighs on us injuriously in the predominance given to letters in education. The question is raised whether, to meet the needs of our modern life, the predominance ought not now to pass from letters to science….. I am going to ask whether the present movement for ousting letters from their old predominance in education, and for transferring the predominance in education to the natural sciences, whether this brisk and flourishing movement ought to prevail, and whether it is likely that in the end it really will prevail. [Matthew Arnold *Literature and science*]

At that time the British Education Department was spreading a system of evening class instruction from which the organized science schools of the next decade were developed. The classes ran through the winter and were examined in May and the teacher received pay according to his results, a pound or two pounds or four pounds for every pass, according to its class and grade. Byatt, who was a university M.A., was considered qualified to conduct classes and earn grants in any of the thirty odd subjects scheduled by the Department, and in addition to his day-time teaching, he was already running evening classes in freehand, perspective and geometrical drawing and in electricity and magnetism
[H.G. Wells *Experiment in autobiography*]

It's time you and Trevor tried to find Newton's rings,' I said. It was a difficult experiment, appropriate as Frank had won a scholarship to Oxford ….. 'Fred,' I said, 'you and Benny can do Kater's pendulum.' This was where my guile came in. The experiment necessitated one of them counting the oscillations of a pendulum and the other watching a clock, so precluding all foolish conversation.
[William Cooper *Scenes from a provincial life*]

In Chaucer's day the syllabus was divided into two parts. The trivium - grammar, logic and rhetoric - was taught first, and provided the basis for the more advanced quadrivium - geometry, arithmetic, astronomy, and music.(Astronomy was regarded as applied geometry, and music as applied arithmetic.) The teaching on these courses relied heavily on the writings of Aristotle. These filtered into Western Europe in the twelfth century, and, by the end of the following century, a Christianised version of the Aristotelian world-picture had come to be widely accepted. So, in the fourteenth century, what Chaucer's Oxford scholar wanted above all else was a set of Aristotle's writings [first extract]. Chaucer was well-grounded in astronomy, and wrote a treatise on the astrolabe - the basic astronomical instrument of the period. (The instrument he describes had been built for use at the latitude of Oxford.)

By the eighteenth century, the Aristotelian world-picture had mostly been discarded, though classical works still provided the basis for much traditional education. Newtonian science was often promulgated via other non-traditional channels - more especially, by independent lecturers. Three of the leading practitioners are mentioned in the second extract. John Rowley was reckoned to be the best British instrument-maker in the early eighteenth century. His 'box' was the orrery - a mechanical model of the solar system which he and others used for teaching. John Desaguliers, who had assisted Newton, was another interested in the design of instrumentation. He was one of the early developers of the planetarium, but his interests ranged widely. William Whiston also knew Newton well. He was Newton's successor as Lucasian Professor at Cambridge, but was expelled for heretical views. He subsequently became a popular independent lecturer in London. The significant point about these lecturers, as Young notes, was that, while the universities were for men only, their lectures were attended by mixed audiences.

By the early nineteenth century, science teaching was beginning to be institutionalised. For example, the Royal Institution in London was by now in existence and Humphry Davy was attracting large audiences to his lectures there. His successor, Michael Faraday, was equally effective. Not only were his audiences mixed, but he made a major effort to interest children in science. The third extract comes from one of the most famous lecture series ever: they were the prototype for the Christmas lectures for young people which still run today. The early nineteenth century in Britain also saw the rise of utilitarian philosophy. Though not specifically concerned with science, it was seen - for example, in the writings of J.S. Mill - as applying a scientific approach to ethics and economics. A major criticism of both utilitarianism and science was that they placed far too much emphasis on factual knowledge.

Dickens caricatures this approach in the fourth extract. Matthew Arnold, as an Inspector of schools, had a more basic concern - what kind of teaching provided the best training for the mind [fifth extract]. The argument for a classical education had traditionally been that it provided such training better than other studies. By Arnold's time this was coming under attack. There was a growing belief that a full education should include some science. Indeed, T.H. Huxley claimed that science could by itself provide an adequate training for the mind (though he accepted the value of some humanities teaching). Arnold, urged the opposite viewpoint: that a properly developed classical education could also embrace science.

In the latter part of the nineteenth century, it became clear that scientific and technical education in Britain was lagging behind what was available amongst its leading commercial competitors. One response was to set up a governmental body - the Science and Art Department - which instituted payment-by-results examinations, as Wells explains in the sixth extract. A bright pupil, such as Wells, could cram for a whole range of examinations and earn a considerable sum of money for the teacher. Many teachers, like Wells' Mr. Byatt, also put on evening classes which brought in more money. In the last decades of the century, higher-grade schools began to receive block grants from the Department in return for including science in their syllabus. This allowed them to build science laboratories. Previously, pupils, such as Wells, could pass science examinations without any practical work. In the twentieth century, the laboratories led to a science examination system that included both theoretical and the practical aspects. Cooper, who taught at a grammar school in Leicester, records a typical physics laboratory session in such a school [last extract]. Newton's rings are an optical phenomenon first studied by Isaac Newton. Kater's pendulum was developed by an army officer, Henry Kater, in 1817 to measure the force of gravity in different locations. Both were experiments in the classical physics that dominated school science teaching in the first half of the twentieth century.

Thence with Creed to Gresham College, where I had been by Mr. Povey the last week proposed to be admitted a member; and was this day admitted, by signing a book and being taken by the hand by the President, my Lord Brunkard, and some words of admittance said to me. [Samuel Pepys *Diary 1665*]

By 1740, barely half the Fellows could be counted on to pay their dues, and some were so severely in arrears that the Society's accumulated deficit had risen to £18,000 – a worrying sum for a private body of modest size. Partly to restore the balance sheet, it began taking in members who were distinguished but not terribly scientific. By the end of the century, Fellows included Edward Gibbon, Warren Hastings and even Lord Byron.[Bill Bryson *Seeing further*]

Last, praise we the noble body to which, for the time, we belong,
Ere yet the swift whirl of the atoms has hurried us, ruthless, along,
The British Association—like Leviathan worshipped by Hobbes,
The incarnation of wisdom, built up of our witless nobs,
Which will carry on endless discussions, when I, and probably you,
Have melted in infinite azure—in English, till all is blue.
[James Clerk Maxwell *British Association, notes of the President's Address*]

May 12, 1827. Joseph Smiggers, Esq., P.V.P.M.P.C. [Perpetual Vice-President--Member Pickwick Club], presiding. The following resolutions unanimously agreed to:--'That this Association has heard read, with feelings of unmingled satisfaction, and unqualified approval, the paper communicated by Samuel Pickwick, Esq., G.C.M.P.C. [General Chairman--Member Pickwick Club], entitled "Speculations on the Source of the Hampstead Ponds, with some Observations on the Theory of Tittlebats;" and that this Association does hereby return its warmest thanks to the said Samuel Pickwick, Esq.,G.C.M.P.C., for the same.'
[Charles Dickens *The Pickwick papers*]

The infant science of Bio-geology--the science which treats of the distribution of plants and animals over the globe, and the cause of that distribution. I doubt not that there are many here who know far more about the subject than I; who are far better read than I am in the works of Forbes, Darwin, Wallace, Hooker, Moritz Wagner, and the other illustrious men who have written on it. But I may, perhaps, give a few hints which will be of use to the younger members of this Society, and will point out to them how to get a new relish for the pursuit of field science. [Charles Kingsley *An Address given to the Scientific Society of Winchester, 1871*]

Now nothing could be finer or more beautiful to see
Than the first six months' proceedings of that same Society,
Till Brown of Calaveras brought a lot of fossil bones
That he found within a tunnel near the tenement of Jones. [Bret Harte *The Society upon the Stanislaus*]

"Scientists" they are, and when they emerge to any sort of publicity, "distinguished scientists" and "eminent scientists" and "well-known scientists" is the very least we call them. Certainly both Mr. Bensington and Professor Redwood quite merited any of these terms long before they came upon the marvellous discovery of which this story tells. Mr. Bensington was a Fellow of the Royal Society and a former president of the Chemical Society, and Professor Redwood was Professor of Physiology in the Bond Street College of the London University. [H.G. Wells *The food of the gods*]

….. to say that the R.S. should refuse such men as Layard – were he still alive – Budge or my son Arthur and refer them to the heterogeneous mass that calls itself the British Academy is to my mind little short of an insult. I do not believe that more than a tenth of the R.S. is in favour of restriction.
[Sir John Evans quoted in A.J. Meadows *Science and controversy*]

Scientific societies DISCUSSION

The Royal Society of London was one of the first scientific societies founded. It was established after the restoration of Charles II, from whom it received its 'Royal' prefix. Gresham College was set up in the City of London at the end of the sixteenth century. It served as the home of the Royal Society from when the latter began in 1660 until 1710. The President when Pepys joined was actually Lord Brouncker - not Brunkard - a highly regarded mathematician [first extract]. Pepys, as his diary reveals, was greatly interested in the work carried out by various members of the Society. In its early days, the activities of the Royal Society were often regarded with considerable amusement and even disdain. Thomas Shadwell's play *The virtuoso* [1676] particularly targeted Hooke, while Jonathan Swift's visit to Laputa in *Gulliver's travels* [1726] targeted Newton. Still, as Bill Bryson points out, financial problems were more pressing than satire [second extract]. He goes somewhat astray, however, in thinking that the recruitment of non-scientists was unusual. The Society originally had a wide brief – to support the undertakings, studies, and labours of the ingenious. Byron, for example, took a considerable, if amused, interest in a number of the topics that concerned the Society.

The British Association for the Advancement of Science was set up in the 1830s, in part as a reaction to the Royal Society, which had by now become a rather stodgy body, very much centred on London. Once the Industrial Revolution was under way, much work of interest to science was being done out in the provinces. The British Association was a more go-ahead organisation, which met each year – as it still does – somewhere outside London. Maxwell became a strong supporter of the British Association – or the 'British Ass' as it was often called - later in the century. Along with other leading scientists, he also became a member of its dining club. It was customary at this 'Red Lion Club' for members to produce songs or verses. John Tyndall was President of the Association for its meeting in Belfast in 1874, and gave a notoriously anti-religious address. Maxwell produced the poem quoted here by way of a commentary on Tyndall [third extract].

Just as the Royal Society had aroused derision in its early days, so, too, did the B.A.A.S. Towards the end of the 1830s, Dickens satirized the British Association in the *Mudfog Papers*, which dealt with the activities of the Mudfog Association for the Advancement of Everything. Around the same time he published the *Pickwick Papers*, which provided a humorous view of the activities of a smaller local society [fourth extract]. 'Tittlebat' was a popular name for the stickleback, a fish that was, indeed, the subject of research at that time. The Hampstead Ponds were originally dug as reservoirs to supply water to London.

Local societies were being formed throughout the nineteenth century. Many included science as part of their brief. Charles Kingsley, a warm supporter both of such societies and of scientific fieldwork, became a canon of Chester Cathedral in 1870. When he gave this talk to the Winchester Society, he was already involved in forming a similar society in Chester. The first four names he mentions are of famous British scientists. The fifth – Moritz Wagner – was a German explorer, who had a well-publicised debate with Darwin on the topic of biogeography and the evolution of species.

Local societies also flourished in America during the nineteenth century. Indeed, because of the distances involved, local and regional societies were initially more important than national societies. Calaveras County was established in California in 1850. It was an important gold-mining centre when Harte wrote this poem about a supposed local society. He was not the only one to derive humour from the area. It was also the subject of Mark Twain's story *The Celebrated Jumping Frog of Calaveras County*.

The nineteenth century also saw the growth of specialisation in science. As scientific knowledge increased, so it became more difficult to make significant advances in more than one field. As a result, specialist societies began to appear. The Chemical Society of London (later the Royal Chemical Society) was founded in 1841. Its presidents were – as Wells says – all eminent scientists. At the same time, to be a Fellow of the [by-now reorganised] Royal Society was also increasingly a matter of some prestige. However, the Royal Society decided at the end of the nineteenth century that, unlike most Continental societies. it should restrict membership to the natural sciences only. Sir John Evans, an archaeologist, had been Treasurer of the Royal Society for twenty years, and is here bemoaning the fact that his subject was now being ejected from the Royal Society into the newly created Royal Academy [final extract]. His son, Arthur, was then in the process of making his famous discovery of the Minoan civilisation in Crete.

Philosophers was felt to be too wide and too lofty a term, and was very properly forbidden them by Mr. Coleridge, both in his capacity of philologer and metaphysician; *savans* was rather assuming, besides being French instead of English; some ingenious gentleman proposed that, by analogy with *artist*, they might form *scientist*.[William Whewell *Quarterly Review* 1834]

The knowledge both of the Poet and the Man of Science is pleasure; but the knowledge of the one cleaves to us as a necessary part of our existence, our natural and inalienable inheritance; the other is a personal and individual acquisition, slow to come to us, and by no habitual and direct sympathy connecting us with our fellow-beings.[William Wordsworth *Preface to Lyrical ballads*]

'Of course,' said Mr. Stockton, 'mere science, as science, does not deal with moral right and wrong.'
'No,' said Mr. Saunders, 'for it has shown that right and wrong are terms of a bygone age, connoting altogether false ideas. Mere automata as science shows we are—clockwork machines, wound up by meat and drink—'
'As for that,' broke in Mr. Storks, who had by this time recovered himself—and his weighty voice at once silenced Mr. Saunders, 'I would advise our young friend not to be too confident. We may be automata, or we may not. Science has not yet decided. And upon my word,' he said, striking the table, 'I don't myself care which we are.   [W.H. Mallock *The new republic*]

The antechapel where the statue stood
Of Newton with his prism and silent face.
The marble index of a mind for ever
Voyaging through strange seas of Thought, alone.
[William Wordsworth *The prelude*]

A little dapper man but with shiny elbows
And short keen sight, he lived by measuring things
And died like a recurring decimal
Run off the page, refusing to be curtailed;
Died as they say in harness, still believing
In science, reason, progress.
[Louis Macneice *The kingdom*]

All these have never yet been seen--
But Scientists, who ought to know,
Assure us that is must be so...
Oh! let us never, never doubt
What nobody is sure about!
[Hilaire Belloc *The microbe*]

The Thinking Machine! Perhaps that more nearly described him than all his honorary initials, for he spent week after week, month after month, in the seclusion of his small laboratory from which had gone forth thoughts that staggered scientific associates and deeply stirred the world at large.
[Jacques Futrelle *The problem of Cell 13*]

It was on a dreary night of November that I beheld the accomplishment of my toils. With an anxiety that almost amounted to agony, I collected the instruments of life around me, that I might infuse a spark of being into the lifeless thing that lay at my feet. [Mary Shelley *Frankenstein*]

William Whewell was a North-country lad who became Master of Trinity College, Cambridge. He was an important member of the scientific community in the first half of the nineteenth century - not least because of his wide range of interests, As Sydney Smith said of him: 'science was Whewell's forte, but omniscience his foible'. He here records a discussion at a meeting of the British Association in 1833, where the question arose as to what people who were interested in science should be called [first extract]. Wordsworth, with whom Whewell was acquainted, employed, like many others, the term 'man of science' [second extract], but this was felt to be cumbersome. Coleridge, who was accepted as an expert in the use of words, was appealed to for his aid. An ingenious gentleman (who was almost certainly Whewell himself) suggested 'scientist'. But agreement proved difficult, and the new name 'scientist' was not immediately accepted. For example, Michael Faraday - for whose researches Whewell coined the terms *anode* and *cathode* - preferred to be called a natural philosopher. But Whewell continued to support the word, and, by the latter part of the nineteenth century, 'scientist' had become a common designation.

Wordsworth draws a distinction between the poet and the scientist which, in a more general form, still exists today. Those on the arts side are seen as more concerned with emotions and people; those on the science side with reason and things. (In other words: arts = 'warm'; science = 'cold'.) In the nineteenth century, this supposed difference was underlined by the role that some leading scientists played in attacking religion. W.H. Mallock's satire on contemporary thinkers, *The new republic*, included three of these scientists expounding their unbelief [third extract]. Stockton is John Tyndall, who was for many years professor of physics at the Royal Institution in London. In his presidential address to the British Association in 1874, Tyndall started a world-wide debate by claiming that religion should not be permitted to 'intrude on the region of knowledge, over which it holds no command'. Saunders is W.H. Clifford, who was professor of mathematics at University College London and an ardent atheist. Storks was the most important of them all. He was T.H. Huxley, professor of natural history at South Kensington. Huxley's position was rather more subtle than that of the other two - he coined the word 'agnostic' to describe it - but he was opposed to organised religion, and he was an influential figure in late Victorian Britain.

In the second extract, Wordsworth speaks of the scientist's 'individual acquisition' of knowledge. The same point comes up in his lines on Newton's statue [fourth extract] with the emphasis on the final word 'alone'. This picture of each scientist ploughing a lonely research furrow was not too bad an approximation when Wordsworth was writing. As scientific research expanded, so did the growth of research teams. Individual scientists working by themselves are now quite rare. Yet this remains the popular picture of research. It is expected that new discoveries will be linked to the names of a limited number of researchers - as, for example, when awarding Nobel prizes.

At the same time, it is accepted that any scientist's work is likely to be superseded by subsequent research. Macneice [fifth extract] writes about an ordinary scientist - not one of the greats - whose research may appear in one edition of a textbook, but will disappear in the next. In the 1940s, when this poem was written, it was automatically assumed that the scientist would be a man. Even today, the default assumption in the world at large seems to be that science is primarily a male activity. But the belief in progress, common among Victorian scientists, had by this time come under scrutiny. The experiences of the Second World War suggested to many that the idea of science always aiding progress was open to doubt.

Scientists have, of course, always been susceptible to caricature. One approach is via ridicule. Shadwell's play, *The virtuoso*, was already poking fun at scientists in the seventeenth century. Belloc, as an arts-oriented Roman Catholic, could afford to feel superior [sixth extract]. Nevertheless, his assertion that scientists are over-confident in their claims currently forms part of the debate on global warming. Fictional scientists are often depicted as highly eccentric. Futrelle's Professor A.S.F.X. Van Dusen [seventh extract] - from the same epoch as Belloc's scientist - is a good example. In the extreme case, we get the 'mad scientist'. Mary Shelley's hero, Victor Frankenstein, is often quoted as the prototype here, though Shelley, herself, might well have disputed this.

Testacea ! with what store immense
Of beauty to the cultur'd sense,
Your various tint allures;
[Sarah Hoare *Conchology*]

I seriously counsel you to try if you cannot find something new this summer along the coast to which you are going. There is no reason why you should not be as successful as a friend of mine, who, with a very slight smattering of science, and very desultory research, obtained last winter from the Torbay shores three entirely new species, beside several rare animals which had escaped all naturalists since the lynx-eye of Colonel Montagu discerned them forty years ago.
[Charles Kingsley *Glaucus*]

Half a century ago, in many parts of the coast of Devonshire and Cornwall, where the limestone at the water's edge is wrought into crevices and hollows, the tide-line was, like Keats' Grecian vase, "a still unravished bride of quietness." These cups and basins were always full, whether the tide was high or low ….. All this is long over, and done with. The ring of living beauty drawn about our shores was a very thin and fragile one.
[Edmund Gosse *Father and son*]

….. she wheeled her wheel-barrow,
Through streets broad and narrow,
Crying, "Cockles and mussels, alive, alive, oh!"
[James Yorkston *Molly Malone*]

The oyster's a confusing suitor;
It's masc., and fem., and even neuter……
I'd like to be an oyster, say,
In August, June, July or May.
[Ogden Nash *The oyster*]

She wandered back to her rock-pools ….. She bent over watching the anemone's fleshy petals shrink from the touch of her shadow, and she laughed to think they should be so needlessly fearful. The flowing tide trickled noiselessly among the rocks, widening and deepening insidiously her little pools. Helena retreated towards a large cave round the bend. There the water gurgled under the bladder-wrack of the large stones.
[D.H. Lawrence *The trespasser*]

On the bottom we could see long snake-like animals, gray with black markings, with purplish-orange floriate heads like chrysanthemums. They were about three feet long and new to us. Wading in rubber boots, we captured some of them and they proved to be giant synaptids. They were strange and frightening to handle.
[John Steinbeck *The log from the Sea of Cortez*]

"Cyanea!" I cried. "Cyanea! Behold the Lion's Mane!" The strange object at which I pointed did indeed look like a tangled mass torn from the mane of a lion. It lay upon a rocky shelf some three feet under the water, a curious waving, vibrating, hairy creature with streaks of silver among its yellow tresses. It pulsated with a slow, heavy dilation and contraction….. "Here is a book," I said, taking up the little volume, "which first brought light into what might have been forever dark. It is   *Out of Doors*, by the famous observer, J. G. Wood. Wood himself very nearly perished from contact with this vile creature, so he wrote with a very full knowledge.
[Conan Doyle *The adventure of the Lion's Mane*]

There are differing definitions of what constitutes 'the sea shore', but the one used here is the 'inter-tidal zone' [ie. the shore-line between high and low tides]. Although this is a relatively limited area, it is scientifically interesting because of the great diversity of life that it contains. Sarah Hoare, a Quaker teacher in the early nineteenth century, was an early advocate of teaching children about natural history. She particularly valued the biology of the sea shore, and her poem, based on the recently introduced Linnaean classification, described its wonders [first extract]. *Testacea* is actually a broader term than *conchology*. It includes all those invertebrate animals which sport shells, whereas the latter term refers specifically to the study of mollusc shells.

Hoare was a forerunner of the explosion of interest in the sea shore that occurred during the nineteenth century. Charles Kingsley strongly encouraged such studies. His *Glaucus, or the wonders of the shore*, published in 1855, pointed out that, besides learning about marine biology, it was still possible to make new discoveries [second extract]. (Colonel Montagu fought in North America in the latter years of the eighteenth century. However, his fame rests on his contributions to natural history after he retired from the Army. His application of the Linnaean classification, not least to specimens from the sea shore, led to the identification of several new species.) Kingsley's interest in the sea shore had been stimulated by Philip Gosse. Gosse's writings on marine biology were immensely popular among the increasing number of people who took seaside holidays in the nineteenth century. His son, Edmund Gosse, wrote a famous autobiography in the course of which he noted that his father's enthusiasm had led to the devastation of sea shores in Britain [third extract]. Indeed, it took many years for the shores to recover from their Victorian tribulations.

The sea shore has been a source of food since the earliest times. By the nineteenth century, the trade, especially in molluscs, was in full flow. The popular Dublin song 'Molly Malone' comes from that century [fourth extract]. Cockles are a type of salt-water clam, though the name is sometimes applied to other molluscs which have a similar appearance. Mussels are a related group, but whereas cockles have a rounded shell, mussels are more elongated. Molluscs for sale inland were kept alive to ensure their freshness. Oysters are yet another group of edible molluscs, but, unlike the others, are sometimes farmed, rather than collected from the wild. Oysters begin life as males, then pass through a quiescent period, and end as females [fifth extract]. As Nash notes, there is an old saying that you should only eat oysters when there is an 'r' in the month. So the months he mentions are the ones that are, from the oyster's viewpoint, safe. (The rationale is that oysters are more likely to go bad in the summer months.)

Rocky shores often display an abundance of seaweeds. These attracted Lawrence's interest: one of his short poems is entitled *Seaweed*. In the sixth extract, he specifically mentions bladderwrack. This is the commonest type of seaweed found round the coasts of the British Isles. It gets its name from the bubbles [bladders] on its fronds, which help to buoy it up. It was the original source of iodine for medical use. Lawrence also mentions sea anemones, some of the most attractive denizens of rock pools (and called after the flower). Most remain stationary, attached to the bottom, and catch their food via a number of waving tentacles. Synaptids, commonly called sea cucumbers, feed in a similar way - Steinbeck describes their tentacles as like chrysanthemums - but they are mobile creatures [seventh extract]. They feel sticky to handle because of hooked spicules in their skin and they are found in greatest diversity in tropical waters such as the Sea of Cortez (otherwise known as the Gulf of California). The Lion's Mane is actually a jelly fish, so not really a feature of the sea shore. However, they have been found there - as described by Sherlock Holmes [final extract]. It is the largest jellyfish of all with tentacles up to 120 feet long. The sting is unpleasant, but not normally fatal. However, Holmes emphasizes that the victim had a weak heart. The Revd. John Wood was, like his contemporary Gosse, a popular writer on natural history. His book *Out of doors* was published in 1874. He describes his own encounter with this jellyfish in the following way: 'at short intervals sharp pangs shot through the chest .... Then the pulsation of the heart would cease for a time .... Then the lungs would refuse to act'.

Then, we upon our Globe's last verge shall go,
And view the Ocean leaning on the sky:
From thence our rolling Neighbours shall we know,
And on the Lunar world securely pry.
[John Dryden *Annus Mirabilis*]

Is yon our Earth? .. Can it be?
Yon small blue circle, swinging in far ether,
With an inferior circlet near it still,
Which looks like that which lit our earthly night?
[Lord Byron *Cain*]

"Suffer me to finish," he calmly continued. "I have looked at the question in all its bearings, I have
resolutely attacked it, and by incontrovertible calculations I find that a projectile endowed with an initial
velocity of 12,000 yards per second, and aimed at the moon, must necessarily reach it. I have the honour,
my brave colleagues, to propose a trial of this little experiment.".…. An appalling unearthly report followed
instantly, such as can be compared to nothing whatever known, not even to the roar of thunder, or the blast
of volcanic explosions! No words can convey the slightest idea of the terrific sound! An immense
spout of fire shot up from the bowels of the earth as from a crater. The earth heaved up, and with great
difficulty some few spectators obtained a momentary glimpse of the projectile victoriously cleaving the air
in the midst of the fiery vapours! [Jules Verne *From the Earth to the Moon*]

First it is fitting to name pioneers:
Tsiolkovski, Ganswindt, Goddard, Oberth, Esnault-Pelterie,
Ley and others formed the VfR: whence in the war
Von Braun worked at Peenemunde, under the cold Baltic,
And the rockets fell, on Antwerp and London.
[Robert Conquest *To launch the satellites*]

Still, seeing Armstrong's strong leg
float down in creepy silhouette
          that first stark second.
[John Updike *Seven new ways of looking at the Moon*]

Lewis's creatures have been disporting themselves ever since they were created and we see them in their
paradise. It is Mars and Venus – both now, owing to space probes and photography, useless as possible
paradises, like those lost civilizations in Africa. [William Golding *A moving target*]

The first airmaker on their tour was... a dark, crouching bulk on a stony ridge, its intake funnel like the
rearing neck of some archaic monster. They pulled up beside it, slapped down their helmets and went one by
one through the airlock... The airmaker was one of the most complicated machines in existence. A thing
meant to transform the atmosphere of a planet had to be. [Poul Anderson *The big rain*]

Dear Voyager:
          This is to thank you for
The last twelve years, and wishing you, what's more,
Well in your new career in vacant space.
When you next brush a star, the human race
Will be a layer of old sediment.
[John Updike *An open letter to Voyager II*]

The possibility of putting objects and humans into space became a matter for discussion in the seventeenth century. Thus John Wilkins, one of the founders of the Royal Society, wrote a book about travelling to the Moon. Dryden, another early member of the Royal Society, pondered on the view to be had from space [first extract]. He mentions scanning the Earth, the Moon, and its 'rolling neighbours' [the planets]. In the early nineteenth century, Byron imagined Cain being taken on a similar tour [second extract]. The scene he describes concurs with pictures of the Earth taken from space in recent decades. The image of the Earth is dominated by the blue of its oceans, while beside it there is the smaller 'circlet' of the Moon. The only discrepant note is the reference to the ether – a transparent substance that was then thought to pervade the universe.

Later in the century came discussion of how spaceflight might be achieved. Verne, ever striving for a scientific description, explains what velocity an object must attain in order to escape from the Earth's gravity [third extract]. Unfortunately, his method of achieving this leaves something to be desired, for the space capsule is fired from a gun. The initial acceleration required would have been sufficient to crush the passengers out of existence. Moreover, friction with the atmosphere would have created a high temperature in the capsule. Early in the twentieth century, the Russian space enthusiast, Tsiolkovski, was inspired by Verne's book, but also pointed out that the method would not work. He became, instead, a pioneer advocate of using rockets. Robert Conquest lists some of the other pioneers in his tribute to the British Interplanetary Society, itself founded as long ago as 1933 [fourth extract]. Hermann Ganswindt was a German contemporary of Tsiokovski, who also put forward plans for a space vehicle. The other people named all come from the next generation. Robert Goddard (inspired by H.G. Wells in his youth) was a pioneer of rocket propulsion in the USA between the wars. Hermann Oberth, an Austro-Hungarian (inspired by Jules Verne), worked on rockets in Germany at the same time along with Willy Ley; as did Esnault-Pelterie, a pioneer aviator, in France. The Verein für Raumschiffahrt [VfR] was formed as a focus for space enthusiasts in 1927. Wernher von Braun joined the VfR in 1930 and went on to lead the German rocket programme in the Second World War. Post-war he was acquired by the United States to help develop satellite launchers. The early satellite launches led on to manned space flight: first round the Earth, and then to the Moon. Neil Armstrong was the commander of the Apollo 11 spacecraft. He became the first person to step onto the Moon's surface in 1969 [fifth extract]. As he did so he said: 'That's one small step for [a] man: one giant leap for mankind'. The event was covered worldwide, and Updike wrote this poem on the same day.

Meanwhile, both the USA and the USSR had started sending a series of probes to our neighbouring planets. The first landing on Venus was achieved by a Soviet spacecraft in 1970. The first successful landing on Mars was made a few years later by a US spacecraft. Just before and during the Second World War, C.S. Lewis wrote his space trilogy. One book was set on Mars and another on Venus. For these books, Lewis relied on the rather limited knowledge of these planets gained from ground-based observation. In the event, the space probes showed that the observations, especially of Venus, had gone astray. Lewis had adapted a proposal that Venus was covered with oceans, whereas the space probes showed that the planet was far too hot to hold liquid water. This is the point behind Golding's remarks [sixth extract].In the seventeenth century, it had been speculated that all the planets harboured life. The space probes of the twentieth century suggested, on the contrary, that maybe none apart from Earth was biologically active. Living on other planets was clearly not going to be easy. This led to the idea of 'terraforming' – of changing a planet, and particularly its atmosphere, in such a way that the planet could be inhabited by terrestrial life. This possibility became a staple of science fiction [seventh extract].

The idea behind space probes is to explore the nature of the solar system. Because of the distances involved, the probes may take years to reach their targets. Exploration beyond the solar system would take many human lifetimes. In 1977, the USA launched two spacecraft – Voyagers I and II – to explore the outer planets. After passing their planetary targets, the two space probes continued onwards, and are now on the boundary between the solar system and interstellar space [final extract]. They should soon be relaying direct measurements of interstellar conditions.

Out on the lawn I lie in bed,
Vega conspicuous overhead
In the windless nights of June
[W.H. Auden *A summer night*]

The great Overdog,
That heavenly beast
With a star in one eye,
Gives a leap in the east.
[Robert Frost *Canis Major*]

All that I know
Of a certain star,
Is, it can throw
(Like the angled spar)
Now a dart of red,
Now a dart of blue,
[Robert Browning *My star*]

But in a tangent off he went
To double stars. Now that
Was most suggestive, so content
And quite absorbed I sat.
[Esther B. Tiffany *Applied Astronomy*]

Many a night I saw the Pleiads, rising through the mellow shade,
Glitter like a swarm of fire-flies tangled in a silver braid.
[Alfred Tennyson *Locksley Hall*]

Holes, punched in the sky, which excited me partly because
Of their Latin names and partly because I had read in the textbooks
How very far off they were, it seemed their light
Had left them (some at least) long years before I was.
[Louis MacNiece *Star-gazer*]

Twinkle, twinkle, little star,
I don't wonder what you are:
For, by spectroscopic ken,
I know you are hydrogen. [Anon.]

….. it says, "I burn."
But say with what degree of heat.
Talk Fahrenheit, talk Centigrade.
Use language we can comprehend.
Tell us what elements you blend.
[Robert Frost *Choose something like a star*]

One of the most obvious features of the night sky is the differing brightnesses of the stars. How bright a star appears to be depends on two factors: its intrinsic brightness and its distance from us. The brightest stars are given their own special names - such as Vega [first extract] or Sirius [second extract]. These two stars dominate their constellations: Vega in the constellation Lyra [the harp], and Sirius in Canis Major [the greater dog]. Stars differ, less obviously, in terms of their colour. Our Sun is yellow. Stars hotter than the Sun have a bluish tinge, whilst those cooler are reddish. Both Vega and Sirius are hotter, and therefore bluer, than the Sun. But it is the positions of these two stars in the sky, rather than their physical properties, that particularly interest Auden and Frost. As the Earth circles the Sun during the course of a year, so the constellations that are easily visible at night change. Lyra is a summer constellation which passes almost overhead at our latitudes. This makes Vega a conspicuous object to anyone looking vertically upwards in mid-year as Auden was. Canis Major, on the contrary, is a winter constellation. Moreover, it never rises to a great altitude in our latitudes, but stays nearer to the horizon. So Frost tells us that the 'great Overdog' [Canis Major] with 'a star in one eye' [Sirius] 'gives a leap in the east' [is rising above the eastern horizon]. Out in space, stars shine steadily, but on Earth they seem to twinkle. This is because their light has to pass through the ever-moving atmosphere of the Earth. The atmosphere bends the starlight by an amount that is continually varying. Not only this, but the refraction works differently for different colours of light. The effect of the atmosphere is greater for stars near the horizon. Consequently, Sirius - which is probably the star that Browning had in mind [third extract] - not only twinkles, it also gives flashes of blue light and of red light.

Stars are scattered all across the sky, with the result that some, by chance, appear to be close together. Sometimes this closeness is not due to chance. The two stars actually are close together and their mutual gravitational attraction causes them to orbit each other. Such linked stars are known as 'double stars'. Double stars became a major research topic in astronomy during the nineteenth century. This was sufficiently widely reported that Esther Tiffany [fourth extract], an American writer in the latter part of the nineteenth century, could use it as a simile for human attraction. On a larger scale, stars can exist together in groups ranging from a few stars to several hundred thousand. These groups are called 'star clusters'. Tennyson [fifth extract] mentions the Pleiades star cluster in the constellation of Taurus [the bull]. The Pleiades is a conspicuous object because it is both nearby and because it contains some intrinsically bright, hot stars. The group has attracted attention from long ago in human history. Its alternative name in English - the 'seven sisters' - derives from Greek mythology. Tennyson's 'silver braid' indicates his knowledge of the telescopic appearance of the Pleiades. Its bright stars are surrounded by a blue haze. This is because the starlight is being reflected from dust particles in an interstellar cloud through which the cluster is currently ploughing. The distances to stars can be measured in terms of the length of time it takes their light to reach us. For Sirius, this is just under nine years, so it is said to be nearly 9 light-years from us. The corresponding figure for Vega is 25 light-years, and for the Pleiades, it is around 425 light-years. So this latter is one conspicuous object whose light certainly started on its journey long before MacNiece saw it [sixth extract].

The best-known literary reference to twinkling stars is surely the children's poem that begins 'Twinkle, twinkle, little star'. This early 19th-century poem by Jane Taylor has been parodied a number of times since - most memorably by Lewis Carroll's Mad Hatter. The seventh quotation is an astronomical parody. Until the nineteenth century, there seemed to be no possible way of determining the chemical composition of stars. The simplest assumption was to suppose that their chemical composition was similar to that of the Earth (containing elements such as silicon, oxygen and iron).As the science of spectroscopy developed, however, it was found that certain wavelengths of light were missing from stellar spectra. This was due to the presence of elements in the stars' atmospheres which absorbed light at those wavelengths. Comparison with spectra in the laboratory then made it possible to identify the elements present in the star. The surprising result - as the parody says - is that the dominant element in the majority of stars is actually the lightest element of all - hydrogen. Deriving the composition of a star in this way is not, however, straightforward. It encounters a problem reflected in the final extract from a poem by Robert Frost. The amount of light an element absorbs depends on the temperature. Stars can differ widely in their surface temperatures. Consequently spectroscopic differences between stars are primarily due to differences in temperature. This effect has to be weeded out before their real 'blend of elements' can be determined.

And therefore is the glorious planet Sol
In noble eminence enthron'd and spher'd
Amidst the other.
[William Shakespeare *Troilus and Cressida*]

I conclude, therefore, and say, there is no happiness under (or, as Copernicus will have it, above) the Sun.
[Thomas Browne *Religio medico*]

With a common impulse the multitude rose slowly up and stared into the sky. I followed their eyes, as sure as guns, there was my eclipse beginning! The life went boiling through my veins; I was a new man! The rim of black spread slowly into the sun's disk, my heart beat higher and higher, and still the assemblage and the priest stared into the sky, motionless.
[Mark Twain *A Connecticut Yankee in King Arthur's Court*]

Direct the telescope upon the sun as if you were going to observe that body. Having focused and steadied it, expose a flat white sheet of paper about a foot from the concave lens; upon this will fall a circular image of the sun's disk, with all the spots that are on it arranged and disposed with exactly the same symmetry as in the sun. The more the paper is moved away from the tube, the larger this image will become, and the better the spots will be depicted. Thus they will be seen without damage to the eye
[Galileo Galilei *Istoria e Dimostrazioni intorno alle Macchie Solari*]

Cultivate a work-lust
That imagines its haven like your hands at night
Dreaming the sun in the sunspot of a breast.
[Seamus Heaney *Station Island*]

Many philosophers imagine that the elements themselves may be in time exhausted; that the Sun, by shining long, will effuse all its light; and that, by the continual waste of aqueous particles, the whole Earth will become a sandy desert. I would not advise any readers to disturb themselves by contriving how they should live without light or water. For the days of universal thirst and perpetual darkness are at a great distance. The ocean and the Sun will last our time, and we may leave posterity to shift for themselves.
[Samuel Johnson *The Idler No.3*]

At last, some time before I stopped, the sun, red and very large, halted motionless upon the horizon, a vast dome glowing with a dull heat, and now and then suffering a momentary extinction. At one time it had for a little while glowed more brilliantly again, but it speedily reverted to its sullen red heat. I perceived by this slowing down of its rising and setting that the work of the tidal drag was done. The earth had come to rest with one face to the sun, even as in our own time the moon faces the earth.
[H.G. Wells *The time machine*]

….. the older boy
goes on about what light years
are, and solar winds, black holes,
and how the sun is cooling
and what will happen to
them all when it is cold.
[Alan Shapiro *Astronomy lesson*]

Sun   DISCUSSION

In the geocentric picture of the solar system, the Sun was essentially another planet, revolving, like them, round the central Earth. But it was always obvious that the Sun was much more important for life on Earth than any of the other bodies; it therefore deserved special attention. Shakespeare [first extract] emphasizes that the Sun is distinctive not only in terms of its size and brightness, but also by its position among the planets. In the geocentric picture, the order of the planets in terms of their distance from the Earth was determined by how fast they appeared to move across the sky. The planet Sol [the Sun] was enthroned amongst the other planets because it lay in their midst. Three planets - the Moon, Mercury and Venus - were thought to be closer to the Earth than the Sun, and three - Mars, Jupiter and Saturn - further away.

By Shakespeare's time, the geocentric system was under attack. Half-a-century before, Copernicus had proposed a heliocentric model, with the Earth moving round the Sun. By the early seventeenth century, this new picture was beginning to displace the old one. Sir Thomas Browne, living in this later period, gave wary acknowledgement to the change [second extract]. The interesting point here is his use of the words 'under' and 'above'. In the geocentric model, there was an absolute up and down: you went up to heaven above the stars, and down to hell in the centre of the Earth. One of the problems posed by the heliocentric model was that it ruined this simple picture.

One of the most impressive astronomical events is when the Moon totally eclipses the Sun. Astrologically, this was not a good thing. Milton reflected this belief in the mid-seventeenth century: '[the Sun] from behind the Moon/In dim eclips, disastrous twilight sheds' [*Paradise lost*]. In Milton's time, such eclipses could not be predicted; by the nineteenth century, prediction had become possible. This allowed intrepid travellers - Tintin, for example, in the twentieth century - to save their lives by predicting eclipses, and so terrifying the local population. Mark Twain's hero provides one of the earliest examples of this [third extract]. Unfortunately, Twain's knowledge of eclipses seems to have been limited. For example, it is not usually possible to see the Moon's disc encroaching on the Sun until near the end, because the Sun is so bright.

Galileo's invention of the astronomical telescope in the early seventeenth century was a major factor in the overthrow of the geocentric picture. Observing the Sun was much more difficult than observing the Moon and planets because of the Sun's brightness - as he explains in the fourth extract. His method of projecting the solar image led to his discovery of sunspots. This significantly opposed the traditional model, which asserted that the Sun was a perfect sphere of light. Spots were clearly blemishes, so making it imperfect. The idea of sunspots as blemishes on the Sun's surface can still be found [fifth extract].

In the old geocentric model, the planets, including the Sun were believed to be changeless. This was no longer true in the new astronomy. By the early eighteenth century, Newton had provided evidence for supposing that light consisted of particles. The highly luminous Sun must therefore be losing vast numbers of light particles all the time. Did this imply that, given enough time, the Sun would eventually lose all its substance and die? Samuel Johnson, in the mid-century, seems to have accepted the possibility of this happening. However, as the fifth extract illustrates, he felt that there were more important things to worry about.

By the end of the nineteenth century, light was seen as consisting of waves rather than particles, but the question of the Sun's future evolution remained. It was now accepted that Sun must have a source of energy to make it bright, and this would eventually all be used up. Wells' time traveller took his machine into the far distant future. The sixth extract reports what he saw. The interesting thing is his picture of the Sun. One of Wells' lecturers at South Kensington, Norman Lockyer, had suggested how stars might change with time. They started, he said, large and red, got brighter and bluer, and ended small and red. Yet Wells unexpectedly describes the aged Sun as 'red and very large'. The final extract poses a similar problem. According to modern theory, the Sun is currently heating up (though it will, indeed, cool down in the dim and distant future). The solar wind is a twentieth-century discovery. The Sun is emitting a continual stream of particles, which flows out past the planets to interstellar space. This stream of material particles may seem to raise the same question as Newton's light particles. However, the loss is small compared with the total mass of the Sun, so it is not currently affecting the way the Sun evolves.

….. one of the King's surveyors was questioning me consarning all the region hereabouts. He had heard that there was a lake in this quarter, and had got some general notions about it, such as that there is water and hills; but how much of either, he knowed no more than you know of the Mohawk tongue. I didn't open the trap any wider than was necessary, giving him but poor encouragement
[James Fenimore Cooper *The last of the Mohicans*]

Lord, I discovered when I discovered love
That day a continent within the mind,
Unstable on the sea, boundaries unlined
Which now I slowly take the measure of.
The coast's determined, the mountains do not move;
Natural harbors and clear springs I find
[Elizabeth Bishop *Washington as a surveyor*]

… the Doctor, having secured a pot of Ale, approaches the Geometers. "Come and meet Mr Tallihoe of Virginia", who proves to be anxious that they visit with Col. Washington, of that Province. "You'll want to have a chat, - he's been out there, knows the country, the Inhabitants, - Surveyor, like yourselves.
[Thomas Pynchon *Mason & Dixon*]

THE work undertaken by the Commission was a triangulation for the purpose of measuring an arc of meridian. Now the direct measurement of one or more degrees by means of metal rods would be impracticable. In no part of the world is there a region so vast and unbroken as to admit of so delicate an operation. Happily, there is an easier way of proceeding by dividing the region through which the meridian passes into a number of imaginary triangles, whose solution is comparatively easy.
[Jules Verne *Measuring a meridian*]

I was taking compass bearings for the Ordnance Survey
On an army training camp on Salisbury plain,
I had packed up my theodolite, was calling it a day,
When I heard a voice that sang a sad refrain
[Flanders & Swann *The armadillo*]

I profess to be a scientific man, and was exceedingly anxious to obtain accurate measurements of her shape; but there was a difficulty in doing this. I do not have a word of Hottentot ….. Of a sudden my eye fell upon my sextant; the bright thought struck me, and I took a series of observations upon her figure in every direction …. this being done, I boldly pulled out my measuring tape, and measured the distance from where I was to the place where she stood, and having thus both base and angles, I worked out the results by trigonometry and logarithms.
[Francis Galton *The narrative of an explorer in tropical South Africa*]

'The books I do not want. Besides, they are logarithms - Survey, I suppose.' He laid them aside. 'The letters I do not understand, but Colonel Creighton will. They must all be kept. The maps - they draw better maps than me - of course…… He fingered a superb prismatic compass and the shiny top of a theodolite.
[Rudyard Kipling *Kim*]

Surveying has a very long history. In England, the first large-scale survey was carried out in the eleventh century to provide input to the Domesday Book. Surveying, in the sense of land measurement, was then very inaccurate, but the term also covered additional duties, such as assessing local resources. Surveying was an activity of national importance, and so, as with the Domesday Book, it often involved the Crown. People labeled *King's surveyors* were commissioned to carry out the work, first in the British Isles and then in overseas territories as they were acquired. The first extract is set in North America in the eighteenth century, where a local trapper is talking to the hero of the novel. Part of the reason for misleading the surveyor was this question of assessing local resources.

In a newly settled country, such as North America, determining boundaries between estates was an essential activity. It was in the mid-eighteenth century that George Washington established his name and reputation by working as a surveyor. He obtained his skills in what was then a customary way: firstly by self-tuition and then by working with established surveyors. In Virginia then surveyors were both very socially acceptable and well paid. The poem by Elizabeth Bishop [second abstract] - only published after her death - sees Washington as thinking of life and love in terms of his profession.

Like most surveyors, Washington was concerned with relatively small areas of land. In the eighteenth century, there was also an interest in much larger surveys to try and determine the shape of the Earth. Newton had predicted theoretically that the Earth is not a perfect sphere. This meant that a degree of latitude would cover different lengths of the surface at different latitudes. Surveys to test this were carried out at a number of places round the world. In the 1760s Charles Mason and Jeremiah Dixon carried out such a measurement in North America. There main purpose, however, was to survey the disputed boundary line between Pennsylvania and Maryland [the Mason-Dixon line]. This also involved working in Virginia, which is where the third extract comes in. The 'Doctor' is Benjamin Franklin, whom the surveyors met. ('Geometers' was an up-market name for surveyors, since 'geometry' means 'earth measurement'.) A Virginian might well think of Washington if the topic of surveying came up, and would think of him as 'colonel' because he was a lieutenant colonel in the Virginia Regiment.

Measuring an extended north-south line - an 'arc of meridian' - provides both information on the shape of the Earth and a basic surveying datum from which other measurements can be made. Verne used such activity in Africa as the plot for one of his books [fourth extract]. The basic surveying technique of triangulation, which involves the measurement of angles, had been developed in the seventeenth century. The theodolite, an instrument that could be used for this purpose, dates from the same general period. But it was only towards the end of the eighteenth century that a really accurate theodolite was developed. This was built in England on a commission from the Ordnance Survey. The Survey was set up as a body for military mapping in the eighteenth century. Its primary task between the 1780s and the 1850s was to map the British Isles, setting up in the process a network of accurately surveyed locations (the 'trig. points'). Revision surveys continued on through the twentieth century [fifth extract].

Surveying was an essential component of most nineteenth-century explorations. Francis Galton, a cousin of Charles Darwin, established his scientific reputation in the mid-century by carrying out a survey of South-West Africa. He subsequently put together a set of notes on surveying that were adopted by the Royal Geographical Society for the guidance of explorers. One of Galton's many interests was anthropology, and the sixth extract illustrates his ingenuity in making relevant observations - in this case of the shape of a local woman. His ability to improvise is reflected in his advice to travellers, as given in his book, *The art of travel; or, shifts and contrivances available in wild countries*, which became a standard source of reference.

One of the great enterprises of the nineteenth century was the Survey of India. By the end of the century, members of the Survey had established a triangulation network throughout the country. Kipling used this as the background for *Kim* [final extract]. Several characters in the novel - including Colonel Creighton - have been identified as based on people whom Kipling knew. Kipling saw the Survey as an excellent cover for a spy network. Though it is true that the surveyors were expected to keep their eyes open (much like the King's surveyors in the first extract), the spy system was probably less well-organised in the nineteenth century than Kipling implies.

What be these two shapes high over the sacred fountain,
Taller than all the Muses, and huger than all the mountain?
.... pass on! the sight confuses -
These are Astronomy and Geology, terrible Muses!
[Alfred Tennyson *Parnassus*]

Now there is scarcely any land hitherto examined in Europe, Northern Asia, or North America, which has not been raised from the bosom of the deep, since the origin of the carboniferous rocks, or which, if previously raised, has not subsequently acquired additional altitude..... The organic remains of these rocks consist principally of marine shells, corals, and the teeth and bones of fish; and their nature, as well as the continuity of the calcareous beds of homogeneous mineral composition, concur to prove that the whole series was formed in a deep and expansive ocean.
[Charles Lyell *Principles of Geology*]

There rolls the deep where grew the tree,
Oh earth, what changes hast thou seen!
There where the long street roars, hath been
The stillness of the central sea.
[Alfred Tennyson *In Memoriam*]

So careful of the type she seems,
So careless of the single life
So careful of the type?', but no,
From scarped cliff and quarried stone
She cries, 'A thousand types are gone:
I care for nothing, all shall go.
[Alfred Tennyson *In Memoriam*]

Some of his remarks still linger fresh in my memory. One night when the moon's terminator swept across the broken ground round Tycho he said, "What a splendid Hell that would make." Again after showing him the clusters in Hercules and Perseus he remarked musingly, "I cannot think much of the county families after that".
[Norman Lockyer *Tennyson as a student and poet of nature*]

Must my day be dark by reason, O ye Heavens, of your boundless nights,
Rush of Suns, and roll of systems, and your fiery clash of meteorites?
[Alfred Tennyson *God and the Universe*]

The world was once a fluid haze of light,
Till toward the centre set the starry tides,
And eddied into suns, that wheeling cast
The planets: then the monster, then the man
[Alfred Tennyson *The Princess*]

Tennyson was profoundly interested in science from his youth. At one stage, he set himself a rigorous programme of study; nearly half of the time was devoted to various branches of science. (He studied chemistry on Tuesday, botany on Wednesday, electricity on Thursday, physiology on Friday, and mechanics on Saturday.) In his adult life, he mixed widely with the leading scientists of his day and learnt a good deal from them. They, in turn, regarded him as a great supporter of science. After Tennyson's death in 1892, Thomas Huxley noted that: 'He was the only modern poet, in fact the only poet since the time of Lucretius, who has taken the trouble to understand the work and tendency of the men of science'. Indeed, the regard which scientists had for Tennyson led to his election to the Royal Society - a most unusual thing for a poet - in 1865, with Huxley as one of his main supporters.

The interesting thing about Tennyson's youthful study programme is that it did not include the two areas of science which were to have the most impact on his thinking - astronomy and geology. The first quotation - from a late poem - reflects his belief that the advances made in these two fields raised problems when writing the sort of poetry that concerned him.

In the 1830s, Tennyson bought a copy of Charles Lyell's *Principles of Geology*, and much preferred its hypothesis - that change came about by slow evolution - to the catastrophic changes assumed by many scientists in the early nineteenth century. The extract from Lyell's book is followed by a stanza from Tennyson's *In Memoriam*, which reflects the influence that the new geology had on him. His acceptance of evolution in a general sense existed long before the *Origin of Species* appeared in 1859, as the next extract illustrates. When, in 1844, a new (and widely thought to be) scandalous text on evolution was published, Tennyson wrote to his publisher, saying: 'I want you to get me a book which I see advertised .... it seems to contain many speculations with which I have been familiar for years, and on which I have written more than one poem.'

From early on, Tennyson saw the impressiveness of the universe as applying a corrective to trivial social concerns. It is told that, as a lad, he encouraged his brother, who was worried about attending a dinner-party, with the words: 'think of Herschel's great star-patches, and you will soon get over all that'. Later in life, he became acquainted with the astronomer, Norman Lockyer, and looked through the latter's telescopes. The comment on the county families recorded here stems from one of these visits. Lockyer, who became a close friend of Tennyson's, was given to producing controversial ideas. One of these was his 'meteoritic hypothesis', which suggested that nebulae [bright patches of light in the heavens] might be the result of vast streams of meteorites [extra-terrestrial rocks] colliding. The idea soon withered under accumulating evidence, but Tennyson illustrates his awareness of Lockyer's work in the next extract.

The nature of the nebulae was, indeed, one of the great questions of nineteenth-century astronomy. Herschel, as Tennyson knew, believed that they were great patches of stars, but others thought they were patches of shining fluid. They pointed to a proposal by the French mathematician, Laplace, that stars and planetary systems formed by the condensation of diffuse matter in space. Tennyson liked this idea and refers to it in the final extract. It was eventually realised that some patches of light were star systems [galaxies], while others were stretches of diffuse matter, so both views contained an element of truth. The reference to 'monster' reflects nineteenth-century interest in the recently discovered dinosaurs.

What then is time? If no one asks me, I know what it is. If I wish to explain it to him who asks, I do not know.
[St. Augustine *Confessions*]

Absolute, true, and mathematical time, of itself, and from its own nature flows equably without regard to anything external, and by another name is called duration: relative, apparent, and common time, is some sensible and external (whether accurate or unequable) measure of duration by the means of motion, which is commonly used instead of true time; such as an hour, a day, a month, a year.
[Isaac Newton   *Philosophiæ Naturalis Principia Mathematica*]

People like us, who believe in physics, know that the distinction between past, present and future is only a stubbornly persistent illusion.[Albert Einstein *Letter to the family of M.A. Besso*]

Pure Time, Perceptual Time, Tangible Time, Time free of content, context, and running commentary—this is   *my* time and theme. All the rest is numerical symbol or some aspect of Space. The texture of Space is not that of Time, and the piebald four-dimensional sport bred by relativists is a quadruped with one leg replaced by the ghost of a leg.
[Vladimir Nabokov *Ada or Ardor: a family chronicle*]

The significance of the concept 'Time' in contemporary philosophy, and the results of its application to all the complexity of life and artistic expression around us, is the main subject of this essay…… we are scrutinizing, in Book I., the ravages of the doctrine of 'Time.'
[Wyndham Lewis *Time and Western man*]

Time present and time past
Are both perhaps present in time future,
And time future contained in time past.
If all time is eternally present
All time is unredeemable.
[T.S. Eliot *Burnt Norton*]

I saw Eternity the other night,
Like a great ring of pure and endless light,
All calm, as it was bright;
And around beneath it, Time in hours, days, years,
Driv'n by the spheres
[Henry Vaughan *The world*]

For   lo!, the days are   hastening   on,
By prophet bards foretold,
When with the ever-circling years
Comes round the age of gold
[Edmund Sears *It came upon the midnight clear*]

And time, that takes survey of all the world,
Must have a stop.
[William Shakespeare *Henry IV Part 1*]

The nature of time has been a topic of interest ever since rational argument began. Heraclitus [c.500BC], for example, emphasized that passing time was related to change – you never step into the same river twice. The analogy between time and flowing water has persisted down the centuries. As a line in one of the best-known hymns has it: 'time like an ever-rolling stream'. One of the basic queries has always been whether our perception of time is ultimately subjective or objective. St Augustine [c.400AD] tended, if tentatively, to the former view [first extract]. Time, he thought, was not a part of the real world, Instead, it was a consequence of the way in which the mind perceived the real world. Some thirteen centuries later, Newton took the opposite viewpoint [second extract]. Newton believed that time existed independently of any perceiver, and flowed at the same rate throughout the universe. However, this absolute time was not what human beings normally perceived. What they dealt with in their daily life was relative time, which depended on changes in their environment. This Newtonian view dominated science for the next two centuries, before, with Einstein, all time became relative [third extract]. (Einstein once commented: 'The only reason for time is so that everything doesn't happen at once'.)

By and large, literary figures have tended to emphasize the subjective view of time – as, for example such different contemporaries as Shelley and Blake. Thus Shelley held that the passage of time was linked to our thought processes. Though Einstein's work eliminated Newtonian time, it did so by concluding that time could not be considered by itself. Time and space were inextricably linked. This was not to everyone's taste. Nabokov was always fascinated by time and its nature, but still held that time and space could be dealt with independently [fourth extract]. In this, he was influenced by the French philosopher, Henri Bergson, who differentiated between the time that people actually experience and the time with which scientists' deal. Nabokov was writing after the introduction of relativity. Bergson, however, formulated his ideas before the idea of a space-time continuum was accepted. Bergson initially opposed Einstein's relativity theory, and when, after the First World War, he tried to reconcile his approach with Einstein's, the latter had to point out mistakes in Bergson's argument.

Bergson exerted a considerable influence on writers in the early twentieth century. Marcel Proust, for example, attended Bergson's lectures at the Sorbonne in the early 1890s, some years before embarking on *A la recherche du temps perdu*. Wyndham Lewis saw Bergson's emphasis on time, at the expense of space, as pernicious [fifth extract], and also attacked what he saw as an over-emphasis on the subjective. Wyndham Lewis's friend, T.S. Eliot, attended Bergson's lectures at the Sorbonne two decades after Proust, but he, too, was inclined to be critical of Bergson's ideas on time. He accepted rather the Christian viewpoint, which saw humans as tied both to everyday time in this world and to eternity in the next. For Eliot, the interest lay in the interplay between these two [sixth extract].

Vaughan's poem reflects a similar contrast as Eliot's between human exposure to time, on the one hand, and to eternity, on the other [seventh extract]. There are two points worth comment. The first is his reference to the 'spheres'. Vaughan was living at a time when the old geocentric picture of the universe was being replaced by one involving a heliocentric solar system. The older picture is invoked here. In it, each celestial object was carried around the Earth by its own transparent sphere, whose was motion driven by the next sphere out. As the spheres of the stars, the Sun, and the Moon were driven round the Earth, they provided us with ways of measuring the passage of time. The other point is Vaughan's image of eternity as a ring. Time flowing for ever is difficult to visualise: the idea of it flowing round a closed path perhaps makes it a little easier. However, this raises the question whether history repeats itself after each time circuit. According to Greek mythology, the golden age occurred in the past and the world had deteriorated since. However, the idea of cyclical time, with events repeating themselves was commonplace throughout the ancient world – in ancient Hindu culture, for example. Sears [eighth extract] seems to be adapting this concept to a Christian viewpoint. The Jewish and, subsequently, Christian view of time was linear. History did not repeat itself: time flowed from a beginning to an end [final extract]. In terms of current astronomical thinking, there is agreement that there was a creation event, at which time in the Universe started, but, as yet, no indication has been found of a terminal event, at which time will cease.

He wiste it was the eightetethe day
Of April, that is messager to May;
And sey wel that the shadwe of every tree
Was as in lengthe the same quantitee
That was the body erect that caused it.
And therfor by the shadwe he took his wit
That Phebus, which that shoon so clere and brighte,
Degrees was fyve and fourty clombe on highte;
And for that day, as in that latitude,
It was ten of the clokke, he gan conclude
[Geoffrey Chaucer   *Introduction to the Man of Law's Prologue*]

And then he drew a dial from his poke,
And looking with lack-lustre eye,
Says very wisely, 'It is ten o'clock
[William Shakespeare   *As you like it*]

When I was a boy, I had a clock with a pendulum that could be lifted off. I found that the clock went very much faster without the pendulum. If the main purpose of a clock is to go, the clock was the better for losing its pendulum. True, it could no longer tell the time, but that did not matter if one could teach oneself to be indifferent to the passage of time.
[Bertrand Russell Introduction to *Words and Things* by E. Gellner]

My grandfather's clock was too large for the shelf,
So it stood ninety years on the floor;
It was taller by half than the old man himself,
Though it weighed not a pennyweight more.
[Henry Clay Work   *My grandfather's clock*]

An open locker against the after-bulkhead caught his eye: it was the likeliest place for the ship's chronometers; he stepped to look within it, and saw that it had contained the chronometers (there were notes of the rates still in the nest) but that the instruments were gone.
[John Masefield   *The Bird of Dawning*]

There were railway hotels, office-houses, lodging-houses, boarding-houses; railway plans, maps, views, wrappers, bottles, sandwich-boxes, and timetables; railway hackney-coach and cabstands; railway omnibuses, railway streets and buildings, railway hangers-on and parasites, and flatterers out of all calculation. There was even railway time observed in clocks, as if the sun itself had given in.
[Charles Dickens   *Dombey and Son*]

There are four vibrators, the world's exactest clocks;
and these quartz time-pieces that tell
time intervals to other clocks,
these worksless clocks work well;
independently the same, kept in
the 41° Bell
Laboratory time vault.
[Marianne Moore   *Four quartz crystal clocks*]

The Sun is the traditional basis for time measurement, with midday defined as the time when the Sun lies due South of the observer. Other times can be determined from the orientation of shadows east or west of the midday line. Shadows of any objects can be used for this purpose: in the first extract, Chaucer's innkeeper looks at those cast by nearby trees. Though the method is simple in principle, it is less simple in practice. The innkeeper takes us through the steps that are needed. A comparison of an object's height with the length of its shadow gives the altitude of the Sun in the sky. However, the result depends not only on the time of day, but also on the time of year and the latitude. Since most people in Chaucer's day stayed close to home, it was mainly the time of year that mattered.

Sundials can help eliminate this need for mental arithmetic. Most such dials have been relatively large and fixed to a post or a wall. But, from the fifteenth century onwards, small, portable sundials came into use. It is one of these that Touchstone, the jester in *As you like it*, pulls out of his poke [meaning, in this case, probably 'pocket' rather than 'bag']. Touchstone's timepiece may well have been a ring dial, looking a bit like a napkin ring. Such a ring has a hole at some point on its perimeter; when this is pointed at the Sun, the sunlight falls on a scale of hours inscribed opposite on the inner side of the ring. The reading can be adjusted to the appropriate date via a scale on the outside of the ring.

Sundials can obviously not be used for measuring time at night, or in bad weather. Mechanical clocks, which overcame this problem, became increasingly important in Europe from the fourteenth century onwards. The problem was regulating them in such a way that they measured time consistently. The major step forward here came in the seventeenth century. In the early part of the century, Galileo noted that the swing of a pendulum marked equal intervals of time. In the latter part of the century, Huygens applied this when developing the pendulum clock. Time accuracy immediately increased from errors of many minutes a day to errors of a few seconds. The difference is noted by Bertrand Russell in the third extract. The time a pendulum takes to swing depends on its length. In manufacturing a clock, it is clearly convenient to have the time of a swing equal to one second. The corresponding pendulum length is about one metre. To accommodate this, along with the clock itself, requires a fairly tall case. The name 'grandfather clock' only appeared towards the end of the nineteenth century (they were previously known as 'long-case clocks'). It is believed that the new name came from the popular song quoted in the fourth extract.

Time is connected to longitude. Every 15° difference in longitude corresponds to a time difference of one hour. Pendulum clocks do not work at sea, so much effort was put in, especially during the eighteenth century, to develop a clock without a pendulum that was accurate enough to allow good determination of a ship's longitude. Such accurate clocks were called 'chronometers'. As the fifth extract indicates, it was preferable to cross-check the time by having more than one chronometer. No chronometer kept perfect time: what was required was a clock with small time errors that were both known and consistent. Such errors were recorded as the chronometer's rating, and the information was stored with the chronometer, as Masefield says. Even for a small country like England, there is a time difference of some 20 minutes from East to West across the widest part. This was unimportant when communications were slow: the advent of railways and the electric telegraph drastically changed matters. In 1847, the railways decided to use London time as the standard for their timetables throughout the country. This was when Dickens was writing *Dombey and Son*, and he notes 'railway time' as another example of the influence of the new railway system on human life.

There is a limit to how accurately even the best pendulum clock can keep time. Throughout the twentieth century much effort was put into developing new, more accurate types of clock. The first of these was the quartz crystal clock, the prototype of which was constructed in the USA at Bell Telephone Laboratories at the end of the 1920s. The quartz crystal can be stimulated electrically to produce very regular vibrations. Marianne Moore wrote about these clocks a decade later. As she says, such clocks do not have the complex works of a traditional clock. As with traditional chronometers, a number of quartz clocks are run together to allow cross-comparison, and they are kept at constant temperature in a vault.

FAUSTUS: How many heavens or spheres are there?
MEPHISTOPHILIS: Nine; the seven planets, the firmament, and the empyreal heaven.
[Christopher Marlowe *Dr.Faustus*]

.... where, begin
The suburbs of creation? where, the wall
Whose battlements look o'er into the vale
Of non-existence! Nothing's strange abode!
.... from yon arch, that infinite of space,
  With infinite of lucid orbs replete,
  Which set the living firmament on fire
[Edward Young *Night thoughts: Night ninth*]

"LET THERE BE LIGHT!" proclaim'd the ALMIGHTY LORD,
Astonish'd Chaos heard the potent word;--
Through all his realms the kindling Ether runs,
And the mass starts into a million suns;
[Erasmus Darwin *The botanic garden: Part I*]

The vast sun-clusters' gather'd blaze,
World-isles in lonely skies,
Whole heavens within themselves, amaze
Our brief humanities.
[Alfred Tennyson Epilogue]

The universe expands and contracts like a great heart.
It is expanding, the farthest nebulae
Rush with the speed of light into empty space.
It will contract
[Robinson Jeffers *The great explosion*]

Once upon a time, way back in the infinitesimal
First fraction of a second attending our creation.
A tiny drop containing all of it, all energy
And all its guises, burst upon the scene.
Exploding out of nothing into everything
[George Bradley *About Planck time*]

"Dark matter" next, "dark energy"...
Where is this darkness none can see?
An arid stretch of times and spaces
With human life its lone oasis?
[Robert Conquest *Getting On*]

[I] watched the superclusters move on paths
Within this mightiest of structured forms ....
I saw the stars burn out, the galaxies
Disperse and perish
[Wade Wellman *Cosmic endings*]

The classical picture of the universe, as described by Aristotle, was absorbed into the West-European world view some three centuries before Marlowe wrote *Dr. Faustus* [first extract]. This geocentric picture had the Sun, Moon, and planets moving round the Earth, each embedded in its own 'sphere' [ie. transparent shell]. The stars were all contained in a single sphere beyond the planets. Beyond the stars was heaven, itself: counterbalancing hell, which was at the centre of the Earth. Yet this picture was changing, even as Marlowe was writing. In the mid-sixteenth century, Copernicus had advocated a heliocentric picture for the planetary system. By the end of the seventeenth century, this had become the new orthodoxy in Western Europe. Both new observations and new theory suggested that the stars were not confined to one region close to the planets, but were spread out over vast distances. By the eighteenth century, it was widely believed that the stars were distributed over infinite space: there was no boundary to the universe [second extract]. This was, indeed, required by Newtonian physics. At any boundary, there would a gravitational pull inwards, not balanced by any outwards pull, so the universe would collapse.

If the universe started as diffuse matter (the 'ether' of the third extract), gravitational attraction would lead this to condense, forming stars and clusters of stars. This is the picture presented by Erasmus Darwin (though, in his view, gravitation could sometimes be opposed by explosive forces acting outwards). Darwin's contemporary, William Herschel, at the end of the eighteenth century, provided observational backing for this picture. He discovered clusters of stars, some dispersed and some compact, and suggested that the latter had developed from the former as a result of infall due to gravity. He also showed that the Sun was embedded in a flattened disc of stars - the Milky Way - and speculated that space was filled with such discs, or galaxies [fourth extract]. During the nineteenth century, Herschel's ideas about galaxies were increasingly opposed, but not forgotten – as Tennyson's verse indicates [fourth extract].

The idea that galaxies would all tend to move together under gravitation was shown, in the early decades of the twentieth century, to be wrong. Galaxies did occur together in groups, but such groups were moving away from each other - the universe was expanding. One question this raised concerned the future. Gravitational attraction was still at work. Would this gradually slow the expansion down and cause the universe to contract again? The answer clearly depended on the total mass of material in the universe, but this, unfortunately, proved difficult to estimate. The American poet, Robinson Jeffers lived at a time (he died in 1962) when either option - perpetual expansion, or expansion followed by contraction - seemed possible [fifth extract]. In either case, another question had to be tackled. Looking backwards in time, the universe must have begun as a highly condensed ball of matter which, for some reason, exploded - the 'big bang', as it came to be called. The sixth extract provides a brief summary of how expansion is currently thought to have happened way back at the start. 'Planck time' - named after Max Planck, the founder of quantum mechanics - is an infinitesimally small unit of time. Its importance relates to what are now defined as the four fundamental forces in the Universe - gravitation along with electromagnetism and the weak and strong nuclear forces. In the beginning, all these forces were merged, but, after one Planck period, they started to separate, beginning with the gravitational force.

Since the 1990s, the picture of how the universe expands has become more complicated. In the first place, it has become apparent that there is far more material in the universe than was previously imagined. This extra material is called 'dark matter', because it cannot be detected by traditional methods. It might be expected that this would slow down the rate at which the universe expands, but the second recent major discovery has been that this expansion actually seems to be accelerating with time. The unknown force causing this acceleration has been labelled 'dark energy' [seventh extract]. At the same time, it has become increasingly clear that the universe has an essentially hierarchical structure - stars are assembled into galaxies, galaxies form groups, groups of galaxies form superclusters - and all of this structure seems to have been laid down in the early phases of the universal expansion. The discovery of 'dark energy' seems to have resolved the question of the future of the universe. It will continue to expand for ever, while its constituent parts will evolve, die and gradually disintegrate [final extract].

So swiftly had the thing happened that barely a quarter of the people going to and fro in Hyde Park, and Brompton Road, and the Exhibition Road saw anything of the aerial catastrophe. A distant winged shape had appeared above the clustering houses to the south, had fallen and risen, growing larger as it did so; had swooped swiftly down towards the Imperial Institute, a broad spread of flying wings, had swept round in a quarter circle, dashed eastward, and then suddenly sprang vertically into the air. A black object shot out of it, and came spinning downward. A man! Two men clutching each other! They came whirling down, separated as they struck the roof of Students' Club, and bounded off into the green bushes on its southward side.
For perhaps half a minute, the pointed stem of the big machine still pierced vertically upward, the screw spinning desperately. For one brief instant, that yet seemed an age to all who watched, it had hung motionless in mid-air. Then a spout of yellow flame licked up its length from the stern engine, and swift, swifter, swifter, and flaring like a rocket, it rushed down upon the solid mass of masonry which was formerly the Royal College of Science. The big screw of white and gold touched the parapet, and crumpled up like wet linen. Then the blazing spindle-shaped body smashed and splintered, smashing and splintering in its fall, upon the north-westward angle of the building.
But the crash, the flame of blazing paraffin that shot heavenward from the shattered engines of the machine, the crushed horrors that were found in the garden beyond the Students' Club, the masses of yellow parapet and red brick that fell headlong into the roadway, the running to and fro of people like ants in a broken ant-hill, the galloping of fire-engines, the gathering of crowds - all these things do not belong to this story, which was written only to tell how the first of all successful flying-machines was launched and flew.
[H.G.Wells *The Argonauts of the air*]

These tales have been compared with the work of Jules Verne and there was a disposition on the part of literary journalists at one time to call me the English Jules Verne. As a matter of fact there is no literary resemblance whatever between the anticipatory inventions of the great Frenchman and these fantasies.... The interest he invoked was a practical one; he wrote and believed and told that this or that thing could be done, which was not at that time done.... But these stories of mine do not pretend to deal with possible things; they are exercises of the imagination in a quite different field.
[H.G.Wells *The complete science fiction treasury*]

"I always mix you up with the man you admire so much - Jools Werne. And", he added with a sly look," you *do* admire him, don't you?" In a flash I saw the man plain. He was a critic. I knew my duty at once: I must kill him.
[Graves and Lucas *The War of the Wenuses*]

I do not see the possibility of comparison between his [H. G. Wells] work and mine. We do not proceed in the same manner. It occurs to me that his stories do not repose on a very scientific basis. ... I make use of physics. He invents. I go to the moon in a cannon-ball, discharged from a cannon. Here there is no invention. He goes to Mars in an airship, which he constructs of a metal which does not obey the law of gravitation. *Ça c'est très joli* ... but show me this metal. Let him produce it.
[Jules Verne *Interview in 1903*]

H.G. Wells was a bright lad from a relatively poor family. His father was, among other things, a professional cricketer who played for Kent. In the 1880s, Wells won a scholarship to the Normal School of Science in South Kensington. The 'Normal' in the title - taken from the Ecole Normale in Paris - meant that the institution was designed to produce teachers, in this case science teachers. Wells failed to complete his degree, and soon took to writing both fiction and non-fiction. He established himself as an author in the 1890s, most notably in the field of science fiction. Initially, he concentrated on short stories. The first extract here comes from one of these - *The Argonauts of the air* [1895].

The question of powered flight had reached an interesting stage by the end of the nineteenth century. In the early 1890s, Hiram Maxim, an American resident in England, built a huge steam-powered aeroplane to see whether, given sufficient power, a winged machine could lift off the ground. After much experimental work, the wheels of his machine did in fact leave the ground. To this extent, he could claim to have 'flown'. Maxim's machine was the inspiration for Wells' story. As Maxim had realised, an ability to lift off the ground was insufficient for flight. A proper aeroplane was one that could be controlled as it flew. (The machine he built had guard rails on either side to prevent it from lifting too far. So its lack of control did not matter.) Wells' story takes up this question of control, and the end of the story, reproduced here, shows the aeroplane crashing due to lack of an adequate control system. Wells was actually quite pessimistic about the immediate prospects for properly controlled powered flight, not expecting it to happen for some decades. Yet, within ten years of this story appearing, the Wright brothers had made their first flights.

Wells obviously delighted in making his aeroplane crash on the Royal College of Science. (This was the new name for the Normal School of Science - it was rechristened in 1890. The Royal College, in turn, became a constituent part of Imperial College in 1907.) Having been ejected by this institution, Wells must have gained satisfaction from describing its destruction. His especial emphasis on the students' building is also a reflection of his own interests. He had expended his energy happily on student activities while at South Kensington. (Indeed, this may have contributed to his academic failure.) Correspondingly, the Students' Club survives in his story when the main building does not.

Wells great predecessor in the field of science fiction was Jules Verne. Verne had started writing such tales in the 1860s, and was still alive when Wells first ventured into this kind of writing. It was natural for the two to be compared. Wells, however, thought that they wrote quite different types of story, and the second extract outlines his argument. (He might well have added that his stories tended to be pessimistic about the impact of technology, whereas Verne's tended to be optimistic.) Despite Wells' opinion, a number of his contemporaries believed he had obtained his inspiration from Verne. The third extract, from a contemporary parody of *The war of the worlds* [1898], illustrates this point. However, Verne, himself, agreed with Wells, as the final extract shows. It is only fair to add that Verne's more 'scientific' approach had its limitations. For example, use of a cannon to fire a spaceship would simply ensure that its occupants were squashed very flat on the floor. Verne also had his Wells' stories mixed up. The anti-gravitational material he mentions is actually central to the action in *The first men in the Moon* [1901] - a lunar, rather than a Martian space trip.

If we compare land animals in respect to magnitude, with those that take up their abode in the deep, we shall find they will appear contemptible in the comparison. The whale is doubtless the largest animal in creation.
[Oliver Goldsmith *Natural history*]

The aorta of a whale is larger in the bore than the main pipe of the water-works at London Bridge; and the water roaring in its passage through that pipe is inferior, in impetus and velocity, to the blood gushing from the whale's heart.
[William Paley *Natural theology*]

Io! Paean! Io! sing.
To the finny people's king.
Not a mightier whale than this
In the vast Atlantic is;
Not a fatter fish than he,
Flounders round the Polar Sea.
[Charles Lamb *Triumph of the whale*]

All the whales in the wider deeps, hot are they, as they urge
on and on, and dive beneath the icebergs.
The right whales, the sperm-whales, the hammer-heads, the killers
there they blow, there they blow, hot wild white breath out of the sea!
[D.H. Lawrence *Whales weep not*]

But the ear of the whale is full as curious as the eye. If you are an entire stranger to their race, you might hunt over these two heads for hours, and never discover that organ. The ear has no external leaf whatever; and into the hole itself you can hardly insert a quill, so wondrously minute is it. It is lodged a little behind the eye. With respect to their ears, this important difference is to be observed between the sperm whale and the right. While the ear of the former has an external opening, that of the latter is entirely and evenly covered over with a membrane, so as to be quite imperceptible from without.
[Herman Melville *Moby Dick*]

A very large sperm whale was locked in deadly conflict with a cuttle-fish or squid, almost as large as himself, whose interminable tentacles seemed to enlace the whole of his great body. The head of the whale especially seemed a perfect net-work of writhing arms--naturally I suppose, for it appeared as if the whale had the tail part of the mollusc in his jaws, and, in a business-like, methodical way, was sawing through it. By the side of the black columnar head of the whale appeared the head of the great squid, as awful an object as one could well imagine even in a fevered dream. ….. The eyes were very remarkable from their size and blackness, which, contrasted with the livid whiteness of the head, made their appearance all the more striking. They were, at least, a foot in diameter, and, seen under such conditions, looked decidedly eerie and hobgoblin-like. All around the combatants were numerous sharks, like jackals round a lion, ready to share the feast, and apparently assisting in the destruction of the huge cephalopod.
[Frank Bullen *The cruise of the **Cachalot***]

Then whale by whale
Blundering on the rock with its red stain
Crammed our winter cupboards with oil and meat.
[George Mackay Brown *The year of the whale*]

Oliver Goldsmith's name is not usually associated with science; yet the epitaph that Dr. Johnson wrote for him described him as: 'A Poet, Naturalist, and Historian'. Goldsmith's *Natural history* actually provides a good outline of eighteenth-century knowledge of the natural world. As the first extract indicates, it was already recognised that the largest whales are larger than any other creature on Earth. (Indeed, it is thought now that the blue whale is the largest creature that has ever existed.) This vast size is reflected in all its organs, as Paley nicely describes [second extract]. (Paley's *Natural theology* was one of Charles Darwin's favourite books.)

The image of the whale as a large, rather floppy animal has been applied, on occasion, to human beings. In the third abstract, Charles Lamb uses it in a satire on the Prince Regent. (His editors - the Hunt brothers - were subsequently jailed for attacking the Prince Regent too enthusiastically.) But the seas he mentions are of interest. It was known that whales migrated, visiting the Arctic Ocean ['polar sea'] in the summer, and returning through the Atlantic Ocean for the winter. It is now recognised that the Arctic regions are hugely productive in the summer months, and it is this availability of food that attracts both herbivorous and carnivorous whales. As the Sun disappears from the region, the productivity rapidly declines, and the whales leave. D.H. Lawrence, a century later, was similarly aware of the migration of whales to the Arctic [fourth extract]. His interest in whales was wider than Lamb's, though his understanding of them sometimes failed. Thus of the three names he mentions, the right whale feeds on plankton, while the sperm whale is carnivorous; so is the hammerhead, but it is a shark, not a whale. Only the sperm whale can therefore be labelled a killer whale. 'There she blows' was, of course, a standard cry on whaling boats. Whales breathe air as other mammals do. When they exhale through their blowholes after a dive, the water vapour present condenses and can be seen at a considerable distance.

It would be impossible to write on whales and literature without including something from *Moby Dick* [fifth extract] - a book which Lawrence held in high regard. Melville knew about whales at first hand - he had been an ordinary seaman aboard a whaler in 1841/42 - and he records his knowledge in detail. His discussion of a whale's ear is just one example of this. He had read about an albino sperm whale nicknamed 'Mocha Dick', who was believed to have been killed near the end of the 1830s. Mocha Dick had survived several harpooning attempts, and was said to attack ships that annoyed him. It is hardly surprising that Melville comments on the difficulty of finding an earhole in a whale. In more recent years, it has been realised that whales hear primarily through their lower jawbones, rather than through an external ear.

Frank Bullen left school at the age of nine and eventually decided to become a seaman. He spent many years at sea, reaching the rank of chief mate. He wrote several books, but it is the one cited here that made his reputation [sixth extract]. . The history of whaling goes back many centuries: by the nineteenth century, it had become a major industry, both in Western Europe and in North America. The main aim was to collect oil from the whales. By the time Bullen's book appeared in 1906, oil from whales had been supplanted by oil from wells on land. Sperm whales ('cachelot' is an alternative name for them) were particularly prized because the spermaceti in their heads made superb candles, and the ambergris in their intestines was valued by perfume manufacturers. (Melville spends some time discussing this in *Moby Dick*.) The epic struggle that Bullen records was a rare event for observers in the nineteenth century. However, it is now known that squids are the main food of sperm whales. Encounters of this sort are commonplace, but usually occur out of sight below the ocean surface. It has been suggested that the enormous eyes which so impressed him have actually evolved to allow squids to detect whales in the depths of the ocean.

Whales which strand themselves on beaches have been known throughout history: sperm whales seem particularly likely to do this. Beached whales often do so individually, but sometimes whole groups do it together. This may be because they are responding to the distress call of the first one to become stranded. As Brown [final extract] says, such beachings may prove a major bonus for nearby communities.

Fair weather cometh out of the north
[*Job* **37**:22]

The North wind doth blow and we shall have snow
[Anon. (Sixteenth century)]

The "Washoe Zephyr" (Washoe is a pet nickname for Nevada) is a peculiarly Scriptural wind, in that no man knoweth "whence it cometh." That is to say, where it originates. It comes right over the mountains from the West, but when one crosses the ridge he does not find any of it on the other side! It probably is manufactured on the mountaintop for the occasion, and starts from there. It is a pretty regular wind, in the summer-time. Its office-hours are from two in the afternoon till two the next morning.
[Mark Twain *Roughing it*]

The Westerly Wind asserting his sway from the south-west quarter is often like a monarch gone mad, driving forth with wild imprecations the most faithful of his courtiers to shipwreck, disaster, and death.
[Joseph Conrad *The mirror of the sea*]

But the equinoctial gales were blowing out at sea, and the impartial south-west wind, in its flight, would not neglect even the narrow Marshalsea.
[Charles Dickens *Little Dorrit*]

In the harbour, in the island, in the Spanish Seas,
Are the tiny white houses and the orange trees,
And day-long, night-long, the cool and pleasant breeze
Of the steady Trade Winds blowing.
[John Masefield *Trade winds*]

"Nothing between here and the Urals. That's where the wind comes from. Nothing to stop it. Straight across all those countries."
[Julian Barnes *East wind*]

If brush'd from *Russian* Wilds a cutting Gale
Rise not .....
The full-blown *Spring* thro' all her Foliage shrinks,
Into a smutty, wide-dejected Waste.
[James Thomson *The seasons (Spring)*]

I well remember that after the very severe spring in the year 1739-40, summer birds of passage were very rare. They come hither probably with a south east wind, or when it blows between these points; but in that unfavourable year the winds blowed the whole spring and summer through from the opposite quarters.
[Gilbert White *The natural history and antiquities of Selborne*]

Ryman and I then had a technical conversation about the implications of averaging out the different horizontal winds to produce a mean and what was really entailed, philosophically, by classing turbulence as a deviation from this already artificial measure. He said the nature of an eddy was difficult to define precisely because its identity was involved with its context; and that despite the mean's artificiality, eddies could not be specified independently of it.
[Giles Foden *Turbulence*]

Folklore about the weather, not least about the winds, exists all over the world. It often contains generalisations based on observations made over many years, and these may be susceptible to scientific explanations. The first two extracts illustrate the obvious point that such weather law depends on the locality. A North wind in Israel was, and is, mainly beneficial; in the UK, it is much less so. The difference can be explained in terms of the differences in latitude and topography. But there are also similarities. The UK is on the Eastern edge of the Atlantic and Israel is on the Eastern edge of the Mediterranean; so, for both, Westerly winds are liable to bring rain. But alongside winds prevalent countrywide, there are also winds generated by local peculiarities. In the third extract, Mark Twain describes the Washoe Zephyr in America. In general, the winds that blow daily across mountain slopes follow a regular pattern - upslope during the day, and downslope during the night. The Washhoe Zephyr does the opposite. It is, consequently, still studied with interest by meteorologists.

Most inhabitants of the UK believe instinctively that not only their rain, but their weather in general comes from the left. This stems from global, rather than regional wind systems. The Sun's heat at the equator causes the air there to rise and flow towards the poles, while cold air from the poles flows in below to replace it. This simple pattern is disturbed by the Earth's rotation. The result is that the atmosphere breaks up into bands in which the airflows are at an angle to the North-South line. Between 30° and 60° in the Northern hemisphere, the prevailing wind direction is from the South-West. This is the typical direction for winds coming across the Atlantic to the UK, as Conrad and Dickens both note. Dickens refers specifically to 'equinoctial gales'. There has long been a belief that the South-West wind blows particularly strongly around the two equinoxes [in March and September]. Thus we have Shelley referring to the 'wild West wind' as the 'breath of Autumn's being'; or Masefield's 'dirty British coaster ..... butting through the Channel in the mad March days'. Unfortunately, this is a piece of weather law that meteorologists have failed to confirm. South of 30° in the Northern hemisphere, the wind direction is reversed, and winds blow steadily towards the equator from the North-East. These are the Trade winds. In the days of sailing ships, they provided the best way of getting from Europe to the Americas. So Masefield (like Conrad, an experienced sailor) refers to an island in the 'Spanish seas' (which extended from the Canaries to Central America), a typical stopping place for ships sailing westward.

Whereas South-Westerly winds tend to be warm, Easterlies in the UK tend to be cold. The difference - as the seventh extract tells us - is because the latter blow over land that is often cold, rather than over warm ocean. There are various points in the UK where one can make Barnes' claim [seventh extract]; places that is, where looking Eastwards along a line of latitude, you find yourself standing on the highest point between here and the Ural mountains. Cold winds are rarely popular, but they are necessary. As Thomson tells us [eighth abstract], without cold during the winter to kill off pests, the next season's growth may be badly affected. Plants are not the only things to be affected by unexpected changes in the wind direction. As Gilbert White points out in the next extract, bird migration can be affected by continuing adverse winds. Since many of our summer migrants come from Africa, a wind from the North-West quadrant, as White describes, can easily hinder their arrival.

We talk happily of wind directions, but our personal experience of wind is that it also varies rapidly, moving in gusts and swirls. In other words, winds are essentially turbulent. For meteorologists, the problem is that turbulent flow is very difficult to describe theoretically. This is what the hero of Foden's novel (who rejoices in the name of Meadows) is discussing with the eminent scientist, Ryman [final extract]. The latter has developed a theory which makes it possible to predict the effect of turbulent regions on each other, and so help weather forecasting. Meadows wants access to the theory to help predict the weather for the D-Day landings, but Ryman, a pacifist refuses to help him. The novel includes some scientists who were actually involved at the time. Ryman is modelled on L.F. Richardson, who was active both in theoretical meteorology and as a pacifist. The author is himself the son-in-law of a former Director of the British Meteorological Office.

..... the Poet, prompted by this feeling of pleasure, which accompanies him through the whole course of his studies, converses with general nature, with affections akin to those, which, through labour and length of time, the Man of science has raised up in himself, by conversing with those particular parts of nature which are the objects of his studies. The knowledge both of the Poet and the Man of science is pleasure; but the knowledge of the one cleaves to us as a necessary part of our existence, our natural and unalienable inheritance; the other is a personal and individual acquisition, slow to come to us, and by no habitual and direct sympathy connecting us with our fellow-beings. The Man of science seeks truth as a remote and unknown benefactor; he cherishes and loves it in his solitude: the Poet, singing a song in which all human beings join with him, rejoices in the presence of truth as our visible friend and hourly companion ...... The remotest discoveries of the Chemist, the Botanist, or Mineralogist, will be as proper objects of the Poet's art as any upon which it can be employed, if the time should ever come when these things shall be familiar to us, and the relations under which they are contemplated by the followers of these respective sciences shall be manifestly and palpably material to us as enjoying and suffering beings.
[William Wordsworth Preface to *Lyrical Ballads*]

..... To the solid ground
Of nature trusts the Mind that builds for aye.
[Wordsworth *Sonnet 34*]

Physician art thou? one, all eyes,
Philosopher! a fingering slave,
One that would peep and botanise
Upon his mother's grave?
[Wordsworth *A poet's epitaph*]

I have sometimes doubted the justice of Wordsworth's denunciation of the gentleman who would peep and botanise upon his mother's grave. There are obvious objections to the process ; but, after all, would not a botanist of any sensibility be more deeply affected by the flowers whose forms he had studied, and whose beauty he had learnt to appreciate, than the ordinary observer who has no special associations with the objects confounded together under the general name of weeds ?
[Leslie Stephen *Playground of Europe*]

Do not all charms fly
At the mere touch of cold philosophy?
There was an awful rainbow once in heaven:
We know her woof, her texture; she is given
In the dull catalogue of common things.
Philosophy will clip an Angel's wings
[John Keats *Lamia*]

Science! true daughter of Old Time thou art!
Who alterest all things with thy peering eyes.
Why preyest thou thus upon the poet's heart,
Vulture, whose wings are dull realities?
[Edgar Allan Poe *To science*]

It might be expected that the Romantic poets of the early nineteenth century would take a much dimmer view of science than their predecessors in the eighteenth-century age of enlightenment. In practice, their position was rather more ambiguous. In the first extract, Wordsworth is outlining his view of the different way in which the poet and the scientist look at the world around them. He is essentially saying that both the pictures presented are legitimate, but that the poet's is more representative of society at large. If, however, scientific ideas should become widespread among society, then it will be the duty of the poet to reflect them equally well. But this view of science was overlain by another of his central beliefs. For him, man and nature were one, and contemplating nature led to a mystical union with its spirit. This theme made him unhappy with the aspect of science that has been described as exploring the frontiers of knowledge with a magnifying glass. The difference is nicely reflected by the short excerpt from one of his sonnets that comes second. When the journal *Nature* - now one of the leading science journals in the world - was established in 1869, one of its aims was 'to urge the claims of Science to a more general recognition in Education and in Daily Life'. It was decided that this short extract from Wordsworth would provide an appropriate motto for the new journal. But there was a change. The original capitalised 'mind', but not 'nature'. The quotation in the journal capitalised 'nature', but not 'mind'. Had Wordsworth been alive in 1869, he might well have approved of the overall aim of the journal, but would have been less happy with some of the science that it published. It hardly fitted with his holistic view of nature.

Wordsworth's viewpoint is reflected in the third of the quotations. Here he is presenting scientific interest as overpowering human concerns (though his poem is actually a tirade against all specialists, not just scientists). The extract that follows is a comment on this poem. Leslie Stephen was a well-known writer and critic during the latter half of the nineteenth century. He is best remembered as founder-editor of the *Dictionary of National Biography* (or, even more often, as the father of Virginia Woolf and Vanessa Bell). Stephen had a high opinion of Wordsworth's poetry, but believed that his ideas relating both to nature and to childhood were open to question.

The extract following Stephen's is from John Keats' poem *Lamia*. Wordsworth can be counted as a first-generation romantic poet, while Keats was second-generation (though Wordsworth outlived him by nearly thirty years). Keats did not have Wordsworth's holistic view of nature, but he did believe that that the rational side of human beings must be kept in check by their emotional side. *Lamia* is essentially an allegory which presents two opponents - one representing emotional life and the other rational life. In this extract, the philosopher [a word often used then for what we would now call a 'scientist'] has destroyed a cherished illusion by his rational thinking. Keats compares this with scientific prying into the beauty of the rainbow. His conclusion, however, is not that rationality should be dismissed: rather that neither reason nor the emotions are acceptable in isolation. Cold philosophy must be balanced by the warmth of the senses. It should be added that the polymath, Thomas Young, investigated optical phenomena on the basis of a wave theory of light in the early nineteenth century. One of the applications of his work was to the understanding of rainbows: a topic which was therefore current when Keats was writing this poem.

Edgar Allan Poe wrote his sonnet *To science* a few years after Keats wrote *Lamia*. Poe was a part of the American Romantic movement and was well-acquainted with Keats' poems, including *Lamia*. Though he disliked allegory, his comments reflect a viewpoint similar to Keats' own. Yet Poe, too, saw science, when mixed with imagination, as acceptable to a writer. He was a pioneer of science fiction. His story, *The unparalleled adventure of one Hans Pfaall* - published in 1835 - reports on a balloon trip to the Moon, so placing itself in the sequence of imagined lunar voyages from the seventeenth century to the present. This, and other stories by him, acted as an inspiration for Jules Verne (whose earliest S-F books also revolved round balloon voyages).

The Bear, the Boar, and every savage name,
Wild in effect, though in appearance tame,
Lay waste thy woods, destroy thy blissful bower,
And, muzzled though they seem, the mutes devour.
More haughty than the rest, the wolfish race
Appear with belly gaunt and famish'd face
[John Dryden *The hind and the panther*]

When the crude embryo careful Nature breeds,
See how she works. And how her work proceeds;
While through the mass her energy she darts,
To free and swell the complicated parts,
Which only does unravel and untwist
Th'invelop'd limbs, that previous there exist.
[Sir Richard Blackmore *The Creation, Book VI*]

The heart and arteries are hollow muscles, and are therefore endued with power of contraction in
consequence of stimulus, like all other muscular fibres; but, as they have no antagonist muscles, the cavities of the
vessels, which they form, would remain for ever closed, after they have contracted themselves, unless some
extraneous power be applied to again distend them. This extraneous power in respect to the heart is the current
of blood, which is perpetually absorbed by the veins from the various glands and capillaries, and pushed into
the heart by a power probably very similar to that, which raises the sap in vegetables in the spring, which,
according to Dr. Hale's experiment on the stump of a vine, exerted a force equal to a column of water above
twenty feet high.
[Erasmus Darwin *Zoonomia Vol. I*]

Coded neurons stream
Into the backbone. Go toward the pines.
He twitches, bound to the landscape:
Deciding and deciding.
[Peter Straub *Wolf on the plains*]

She went to Capes with that riddle and put it to him very carefully and clearly, and he talked well - he
always talked at some length when she took a difficulty to him - and sent her to a various literature upon the
markings of butterflies, the incomprehensible elaboration and splendour of birds of Paradise and humming-
birds' plumes, the patterning of tigers, and a leopard's spots. He was interesting and inconclusive, and the
original papers to which he referred her were at best only suggestive.
[H.G. Wells *Ann Veronica*]

After a day or two the spawn is laid in long strings which wind themselves in and out of the weeds and soon
become invisible. A few more weeks, and the water is alive with masses of tiny tadpoles which rapidly grow
larger, sprout hind-legs, then forelegs, then shed their tails: and finally, about the middle of the summer, the
new generation of toads, smaller than one's thumb-nail but perfect in every particular, crawl out of the water
to begin the game anew. I mention the spawning of the toads because it is one of the phenomena of spring
which most deeply appeal to me, and because the toad, unlike the skylark and the primrose, has never had
much of a boost from the poets.
[George Orwell *Toads*]

Fables about beasts have been popular for many centuries. Thus, Aesop's fables in Latin were used as teaching material in mediaeval times, while Uncle Remus stories have been popular since Victorian times. They reflect the characteristics that the various animals were widely supposed to have (also described in the mediaeval bestiaries). Dryden's poem [first extract] is a seventeenth-century contribution - in this case, a religious allegory. The hind and the panther in the title are the Roman Catholic church and the Church of England, respectively. Of the beasts mentioned in the extract, the bear represents the Independents and the boar is the Anabaptists. In the context of the times, these two groups were less influential than the Presbyterians [the wolf].

Blackmore was a well-known physician in the decades around 1700. He attended both King William and Queen Anne and his skill was commended by John Locke. He particularly delighted in writing epic verse, which was widely sneered at by his contemporaries, though *The creation* - his best-known poem - received commendation from Addison and, later, Samuel Johnson. The poem has been described as a sort of English, Christianised version of Lucretius' *De rerum natura*. Whatever his merits as a poet, Blackmore was a good physician who stressed the importance of observation and experience in treating patients. The interest of the second extract is that Blackmore is explaining the preformation theory of embryology. This supposes that organisms develop from miniature versions of themselves, in contrast with epigenesis where the form of an organism emerges gradually as the embryo develops. The latter view became dominant after Blackmore's death, and has remained so ever since.

Erasmus Darwin used the word *zoonomia* to mean the laws of animal life [third extract]. By his time, the circulation of blood round the body was firmly established. However, he evidently felt that extra pressure was needed to keep the circulation going. For this, he appealed to the research of Stephen Hales earlier in the eighteenth century. Hales measured what is now called the 'root pressure' exerted by a plant: in essence, how tall a column of water the plant roots can be support. This pressure is due to osmosis - the tendency for water to move from a lower concentration solution to a higher concentration solution (in this case, from the soil to the plant). Hales used a similar technique - narrow tubes inserted into arteries - to measure blood pressure.

It was recognised long ago that the nervous system was the controlling mechanism in animals - including human beings - but it was not until the early twentieth century that the way in which it worked began to be understood. It was then realised that basic units called 'neurons' were responsible for transmitting electrical impulses round the system. Different neurons can be coded in different ways: that is, they respond to different stimuli. Straub's poem is an attempt to see the North American plains from the point of view of a wolf [fourth extract].

Well's Ann Veronica is a young woman trying to live independently in London a century ago. (The novel was published in 1909.) She decides to study biology to give herself a better understanding of the world [fifth extract]. In this extract, she is discussing the question of animal markings with her supervisor, Capes. The evolutionary significance of such markings had been discussed by Darwin in *Origin of species*. It had been a matter for debate in the 1890s when Wells was a student in South Kensington. The institution there that Wells attended became one of the constituent parts of Imperial College in 1907. However, he describes the college that Ann Veronica was attending as the 'Central Imperial College' near Regents Park.

Orwell took an interest in the world of nature from his childhood days. His later writings on natural history - such as the final extract - were criticised by some of his left-wing contemporaries on the grounds that he ought to be concentrating on politics (though John Betjeman encouraged him). In view of Orwell's comment on toads and poets, it is interesting that Philip Larkin employed the toad as a metaphor [for stagnation] in his writings. Orwell, of course, leads us back us the first extract, since he wrote the most important beast fable of the twentieth century - *Animal farm*.